Nietzsche's Earth

Nietzsche's Earth

GREAT EVENTS, GREAT POLITICS

Gary Shapiro

The University of Chicago Press CHICAGO & LONDON

Gary Shapiro is the Tucker-Boatwright Professor Emeritus of Humanities and Philosophy at the University of Richmond. He is the author of many books, including *Earthwards: Robert Smithson and Art after Babel* and *Archaeologies of Vision: Foucault and Nietzsche on Seeing and Saying*.

The University of Chicago Press, Chicago 60637
The University of Chicago Press, Ltd., London

Printed in the United States of America

25 24 23 22 21 20 19 18 17 16 1 2 3 4 5

ISBN-13: 978-0-226-39445-9 (cloth)
ISBN-13: 978-0-226-39459-6 (e-book)
DOI: 10.7208/chicago/9780226394596.001.0001

Library of Congress Cataloging-in-Publication Data
Names: Shapiro, Gary 1941– author.
Title: Nietzsche's Earth : great events, great politics / Gary Shapiro.
Description: Chicago : The University of Chicago Press, 2016. |
Includes bibliographical references and index.
Identifiers: LCCN 2016000008 | ISBN 9780226394459 (cloth : alk. paper) |
ISBN 9780226394596 (e-book)
Subjects: LCSH: Nietzsche, Friedrich Wilhelm, 1844–1900—Political and social views. |
Earth (Planet)—Philosophy. | World politics—Philosophy.
Classification: LCC B3318.E27 S53 2016 | DDC 193—dc23 LC record available at
http://lccn.loc.gov/2016000008

♾ This paper meets the requirements of ANSI/NISO Z39.48-1992 (Permanence of Paper).

For Tanja Softić

CONTENTS

PREFACE

We should listen attentively to Nietzsche's call to be loyal to the earth—Zarathustra's first and signature injunction. This call is best heard and understood together with *Ecce Homo*'s bold, hyperbolic claim: "only with me does the earth know great politics." While Nietzsche's earth is certainly a realm of desire, adventure, and passionate individual affirmation, this captures only part of his meaning when he insists that we should care about its direction and future. The challenge is to ask what the human-earth shall become, not only how each of us individually can best live an earthly life. Above all, Nietzsche urges us not to despair of earth's future, as some both in his time and ours do when they celebrate or mourn the "end of history."

Already in Nietzsche's nineteenth century, many in the chattering classes—he calls them the multitude—were fascinated with both optimistic and pessimistic versions of the narrative launched by Hegel, who spoke of his time as the destined culmination of the world, associating it (at least rhetorically) with the Christian end of days. Now we see cruder, fervent, and frequently violent revivals of apocalyptic religious passions. At the same time, the apparent supremacy of the global market appears to some as a reason for either smug self-congratulation or resigned acceptance of an irreversible obstacle to free human development. Still others see environmental crisis as either threatening the human future or provoking the species to reverse course and make a new peace with nature. *Nietzsche's Earth* aims at triangulating the philosopher's thought between nineteenth-century versions of these ideas and attitudes and those proliferating now. Nietzsche, I argue, is one of the very few major philosophers to have taken on characteristic questions posed by modernity when it first became possible to do so, given the nineteenth-century zenith of the nation-state and the new speeds of industry, transportation, and communication.

There are, notoriously, many Nietzsches. This is not the place to sort, rank, and evaluate them, an enterprise that will no doubt continue as long as scholarship is possible. The Nietzsche encountered here is the one who began to ask what the earth that human beings inhabit, cultivate, and contest might be; this would require freedom from the ideological blinders of the "world-history" that fascinated his contemporaries and still tempts us. When Nietzsche's madman disrupts the everyday marketplace with his performance art piece on the death of God, he says that we have barely begun to understand this news. This book argues that, for Nietzsche, "world-history" and the states, churches, and other institutions it celebrates, are among the most problematic "shadows of God" (as he called them in *Gay Science*). These shadows claim to found, center, and encompass all significant events and exhaust all meaning. The "new idol" of the state, says Zarathustra (and the same would hold of the world market), proclaims that outside its bounds there is no value.

By giving pride of place to the earth, sometimes called the "human-earth," Nietzsche shifts focus, paradigm, and perspective away from "so-called world-history." The world, Nietzsche saw, had been conceived in terms that were ultimately metaphysical and theological, understood as an absolute unity, whether as Hegel's succession of states constituting "God's march through the world," or as a unified, globalized economic marketplace. This last version, anticipating today's neoliberalism, was that of Eduard von Hartmann, one of the most popular philosophical writers of the late nineteenth century and the target of Nietzsche's parodic wit. The one thing on which Nietzsche agrees with Hartmann is that it would not be worth living in such a world.

The seeds of *Nietzsche's Earth* germinated in the recognition that many of the politically oriented topics he discussed are close to those emerging since the sea change marked approximately by the fall of the Berlin wall. Immediately we heard warmed-over versions of the end-of-history story, glorifying either the democratic parliamentary state, the infallible world market, or some hybrid of the two. These were soon followed by generally surprising new movements of peoples, newly intense conflicts fought under religious banners we'd assumed were outdated, and consequent clumsy and destructive reactions by states intent on preserving authority and territory. As some of the Cold War's fog and smoke cleared, it was replaced on almost every continent by the confusions of new wars, including civil wars and the poorly named, preposterous "global war on terror." Threatening to overshadow all this was an immense and pervasive atmospheric disturbance of the earth—global warming or climate change.

Those seeking ways to make sense of these startling and unpredictable events of recent decades might want to see how Nietzsche dealt with related problems in the century before last. We can recontextualize his thought in terms of issues and currents of ideas that bear some resemblance to our own situation. Nietzsche tackled the question of world-history head on: once we explicate its ontological foundations, we can ask whether there really is such a thing. He went on to expose the desperate condition of the state, analyzing such symptoms as "culture wars" and "states of exception" to deal with alleged internal and external threats (both were signature tactics and terms of Bismarck's *Reich*). Such maneuvers, Nietzsche thought, were reactive and deceptive attempts to cover over what was happening on the human-earth. The earth could no longer be contained within a world of "peoples and fatherlands." It was witnessing such developments as a movement of peoples that amounted to a new nomadism, and the growth of a transnational audience for news and sensation. Its sense of itself was based on suppressing memories of such things as religious wars that fail to respect national boundaries; its leaders encouraged industrial and bureaucratic standardization that frustrated attempts to live life on a human scale.

World-history involves a philosophy of time, which it sees unfolding toward a goal, punctuated by "great events" involving the rise and fall of states and the careers of world-historical individuals. Nietzsche challenges this theory and experience of time on several fronts. Great events are not noisy but creep in on doves' feet. Those concerned for the earth's futurity must be vigilant in watching for the imbalanced times when it may be possible to seize the fleeting moment, the *kairos,* or opportunity, as it rushes by. Excessive detachment and balance, the hypertrophy of the historical sense, mere observation, or surrender to the unfolding "world-process" are recipes of passive nihilism that will blind us to rare opportunities.

Do not read this book expecting that Nietzsche will provide a specific plan or program for earth's transformation. He would hardly be a thinker of futurity if he did so. The book may disappoint some because it passes lightly over Nietzsche's varied, often inconsistent speculations about possible political futures. Most of these are as ephemeral and foolish as the notebook jotting where he wishes that Germany would seize Mexico. Perhaps the closest Nietzsche comes to sketching a more specific future for the human-earth is in scattered but incisive thoughts projecting the earth's transformation into a garden or a "great tree of humanity." If these hopes remain rather schematic, they still give us something rich to think with. Contextualized in relation to the almost forgotten intersection of the political and the aesthetic that clustered around the idea of the garden, they can contribute to reconceiving the geoaesthetics of the earth at a time of environmental crisis.

Nietzsche can be read productively as devising a set of concepts for understanding earth and its times, evading the traps set by world-history and its current analogues, traps such as the closure of the future, the fetishism of the state, and the fickle, media-mad taste of the multitude. Crucial to these analyses is the effort to rethink the time—or better, the plural *times*—of the earth. Nietzsche's last substantive book was *The Antichrist*. This deliberately sensational text is more than a mad diatribe. The book attacks the Christian foundations of world-history, its strategies of deferral and mortgaged time; it turns Christianity's own concepts against itself. While calling for a new division and reckoning of chronology, at a more radical level *The Antichrist* is Nietzsche's way of saying that it is later than we think. It is past time for thinking earth's times differently and so opening a space for a great politics of the earth.

ACKNOWLEDGMENTS

Nietzsche's Earth has taken shape gradually over a number of years in various venues, including publications, seminars, and talks. This book that thematizes debt has doubtless acquired more debts than I can record or recall. Some of these will be evident from references in the text. I am especially mindful of the collegial support and constructive criticism I have received from Christa Davis Acampora, Keith Ansell-Pearson, Babette Babich, Debra Bergoffen, Edward Casey, Daniel Conway, Lawrence Hatab, Vanessa Lemm, Paul Loeb, Alexander Nehamas, Marcia Cavalcante Schuback, Herman Siemens, Fredrika Spindler, and Tracy Strong. I am grateful to Robert Gooding-Williams and an anonymous reader for extremely helpful comments on the first version of this manuscript. Elizabeth Branch Dyson has been as patient, wise, and discreet an editor as I could wish for. At the University of Richmond I explored ideas that eventually found their way into this book in courses and seminars with open and responsive students. Participants in a semester-long faculty seminar on *Beyond Good and Evil* at Richmond offered stimulating observations and questions. The staff of Boatwright Library was consistently available and helpful. During my last few years at Richmond, I was especially fortunate to have the unflagging support of the departmental administrative coordinator, Michelle Bedsaul, always cheerful and ready to go the extra mile. Michelle was clearly the most helpful person in an administrative position that I worked with in twenty years at Richmond. Also in Richmond, Reingard Nethersole drew on her great knowledge of philosophy and literature to offer invaluable suggestions and clarifications in many conversations or "salons." In 2011 I taught two intensive courses at Södertorn University in Stockholm, where an international group of students provided new perspectives and questions. Tanja Softić, a true lover of the earth and Diretoressa of the Softić Institute, created an atmosphere that made it possible for this rat nose to complete the project. My deepest thanks are to her.

NIETZSCHE'S WORKS

GERMAN

Sämtliche Werke: Kritische Studienausgabe. Edited by Giorgio Colli and Mazzino Montinari. 15 vols. Berlin: de Gruyter, 1980. (*KSA*)

Sämtliche Briefe: Kritische Studienausgabe. Edited by Giorgio Colli and Mazzino Montinari. 8 vols. Berlin: de Gruyter, 1986. (*KSB*)

R. Oehler and A. Bernoulli, eds. *Nietzsches Briefwechsel mit Overbeck*. Leipzig: Insel-Verlag, 1916.

ENGLISH TRANSLATIONS

Comprehensive and standard:

Complete Works of Friedrich Nietzsche. Edited by Ernst Behler, Bernd Magnus, Alan Schrift et al. Stanford: Stanford University Press, 1999– (*CWFN*)

Other translations used and consulted:

Basic Writings of Nietzsche. Translated by Walter Kaufmann. New York: Modern Library, 2000.

Beyond Good and Evil. Translated by Adrian Del Caro. Stanford, CA: Stanford University Press, 2014.

Beyond Good and Evil. Translated by Judith Norman. Cambridge: Cambridge University Press, 2002.

The Birth of Tragedy and Other Writings. Translated by Ronald Speiers. Cambridge: Cambridge University Press, 1999.

Ecce Homo. Translated by R. J. Hollingdale. New York: Penguin, 1979.

Human, All-Too-Human. Translated by R. J. Hollingdale. Cambridge: Cambridge University Press, 1986.

On the Genealogy of Morality. Translated by Carol Diethe. Cambridge: Cambridge University Press, 2007.

Philosophy in the Tragic Age of the Greeks. Translated by Marianne Cowan. Chicago: Henry Regnery, 1962.

Selected Letters of Friedrich Nietzsche. Translated by Christopher Middleton. Chicago: University of Chicago Press, 1969.

Thus Spoke Zarathustra. Translated by Graham Parkes. New York: Oxford, 2005.

Twilight of the Idols and *The Antichrist*. Translated by R. J. Hollingdale. New York: Penguin, 1971.

Untimely Meditations. Translated by R. J. Hollingdale Cambridge: Cambridge University Press, 1983.

The Will to Power. Translated by R. J. Hollingdale. New York: Vintage, 1968.

Writings from the Late Notebooks. Edited by Rüdiger Bittner. Cambridge: Cambridge University Press, 2003.

Young Nietzsche and Philosophy. Translated by Richard Perkins. Mount Pleasant, MI: Enigma Press, 1978.

KEY TO REFERENCES

Nietzsche: References to all works, other than *KSA* and *KSB*, are to numbered sections and aphorisms or chapters, including occasionally to P for Preface or Prologue. References to *KSA* and *KSB* are to volume and page number, occasionally in the case of *KSB* to letter date.

AC: *The Antichrist*
AOM: *Assorted Opinions and Maxims*
BGE: *Beyond Good and Evil*
BT: *The Birth of Tragedy*
CW: *The Case of Wagner*
D: *Dawn*
EH: *Ecce Homo*
GM: *On the Genealogy of Morality*
GS: *Gay Science*
HAH: *Human, All-Too-Human*
KSA: *Sämtliche Werke: Kritische Studien-Ausgabe*
KSB: *Sämtliche Briefe: Kritische Studienausgabe*
PTG: *Philosophy in the Tragic Age of the Greeks*
T: *Twilight of the Idols*
UO: Unmodern Observations I–IV
WP: *The Will to Power*
WS: *The Wanderer and His Shadow*
YNP: *Young Nietzsche and Philosophy*
Z: *Thus Spoke Zarathustra*, to four numbered parts (I–IV and to chapters within them; P = Prologue)

CHAPTER 1

Introduction: Toward Earth's "Great Politics"

> The time for petty politics is over: the next century will bring the struggle for dominion of the earth (*Erd-Herrschaft*)—the *compulsion* to great politics.
>
> NIETZSCHE, *Beyond Good and Evil*, 208

In *Ecce Homo*, Nietzsche offered a self-consciously extravagant, grandiose, histrionic account of his work and its significance. The extravagance is underscored by the title, with its allusion to Jesus's appearance before Pilate and its concluding question and response: "Have I been understood?—Dionysus versus the Crucified" (*EH* Destiny 9).[1] Later we will find reasons for emphasizing that, for the late Nietzsche, Dionysus is the name of the Antichrist. It is easy to be distracted by the fireworks and "dynamite" of this text. Yet allowing for the hyperbolic and parodic modes of which Nietzsche was notoriously a master, there are exclamations and declarations that have uncannily become more chilling and meaningful with time, especially those speaking of political upheaval and possibility, such as this gloss on "Why I Am a Destiny":

> When truth enters into a fight with the lies of millennia, we shall have upheavals, a convulsion of earthquakes, a moving of mountains and valleys, the like of which has never been dreamed of. The concept of politics will have merged entirely with a war of spirits; all power structures (*Machtgebilde*) of the old society will have been exploded—all of them are based on lies: there will be wars the like of which have never yet been seen on earth. Only beginning with me is there *great politics* on earth (*EH* Destiny 1).

Since Nietzsche wrote these words, finally published a few years before World War I, many have read them as prescient prophecy of the new wars that succeeded the relatively limited wars of European nation-states in the 250 years following the Peace of Westphalia that established the context and ground rules of the old politics and "power structure."[2] Certainly, wars have expanded in scope and consequences. New weapons (biological and aerial), new ideologies promulgated by new media (Nazi radio and jihadist social media), threats of global destruction and environmental devastation, new religious wars, war on civilians, the looming specter of wars for the most elementary resources (such as food and water)—all of these could be seen as realizing Nietzsche's oracular utterance. This book inquires into Nietzsche's conjunction of the rethinking of the political on earth. It is not concerned with the specific lineaments of his future and our present, but with the way in which thinking "great politics on earth" means reconceiving human futurity. What is a great politics of the earth? How can we begin not so much to envision specific futures as to incorporate the always indeterminate futurity of the earth into our concept of the political?

All too many of those who have written about Nietzsche's political thought have easily assimilated it to patterns and concepts with which they were already familiar. They took sides on questions having to do with the state, race, democracy, and other themes, asking just where Nietzsche could be placed within a spectrum of possible positions that, they assumed, had already been mapped out. In some cases (as I'll document later) the assumptions of Anglophone scholars as to what Nietzsche was talking about were so deeply embedded as to lead to mistranslations of some crucial terms in his political vocabulary; albeit unconsciously they created misreadings which were structurally similar, if differing in content, to the notorious distortions by Nazi or proto-Nazi enthusiasts like those promoted by Nietzsche's sister Elizabeth and her associates.

Too much time has been spent attacking or praising Nietzsche's political thought for its supposed affinities with a specific form of polity or regime. In contrast, I argue that we should be attending to his struggle to keep the political future open. We need to understand the prominence of earth in his political thought and its relation to his analysis of the state, temporality, and the residues of political theology. Whether commentators see Nietzsche as attempting to renew the ancient Greek *polis* or the Roman *imperium*, delineating a new form of aristocracy, or providing grounds for anarchism, democracy, or even revolutionary socialism, these efforts, I believe, miss the most radical dimension of the "philosophy of the future" that he preludes in *Beyond Good and Evil.*[3] While some of these Nietzsche readings are important antidotes to reactionary or nostalgic ones that see him simply as spokesman for a revived form of slavery and tyranny, they show, more importantly, that he was capable of thinking of something new and different. This is so even if he sometimes expressed this fancifully, as in his notebook suggestions that Germany should conquer and colonize Mexico or his published wish for intermarriage between the Prussian officer class and wealthy Jews (*KSA* 9.546; *BGE* 251). I mean to concentrate on his thought of futurity in a broadly political context, futurity in the sense of that which has yet to be, of the unknown and unknowable which may arise, the "great event" or "great politics" of the earth that become insistent themes of his later work. Those attuned to futurity are open to seizing the gift of fortune, the moment of opportunity, the fleeting moment of great possibilities that the ancients call *kairos* (Machiavelli's *occasione*). What gives Zarathustra the horrors in the specter of the last human is the foreclosure of futurity. These last humans no longer remember what nobility and distinction are. History has come to an end for them in a regulated alternation of work that is not too onerous and play that never touches the danger zone, their little pleasures for the day and the night.

Responses to Nietzsche's political thinking have been strangely silent or vague about what he consistently describes as the site of the political, the earth. Fidelity to the earth, being true to the earth, willingness to sacrifice oneself for the earth, vigilantly dedicating oneself to the earth's direction or meaning (*Sinn*)—these are the repeated refrains of *Zarathustra*. The true danger of the last humans who securitize themselves against all danger is that they will further shrink the earth, obliterating its opportunities and chances. When Nietzsche has Zarathustra speak of the shrinking earth of the last humans, he thinks not only of the unifying effects of world commerce and communication that Marx and others had already seen and that we now call globalization. More emphatically he voices his fear of the disappearance of open seas and horizons. These promising future horizons, promises one might say of futurity itself, were paradoxically necessary to the very enterprises whose development made the last humans possible, enterprises such as the maritime explorations by Genoese and Venetians that he admired. Hopping about like fleas on the contracted earth, the last humans are oblivious of opportunities for decisive and innovative action that could contribute to the "great event." For these risk-averse creatures all is calculable. They take their measured pleasures and distractions in regular doses, failing to look beyond the amusements and intoxicants of consumer culture in the stabilized state.

Nietzsche's worry is that both the future and futurity of the earth are at stake. From Martin Heidegger to Jacques Derrida, Gilles Deleuze, and Alain Badiou, philosophers since Nietzsche have attempted to think related questions. They struggle, as philosophers must, against the conservatism of any given language by articulating and explicating, unfolding the *pli* or fold, in terms such as *Er-eignis* or *l'a-venir* in order to evoke a sense of the futural or evental. These thinkers agree with Nietzsche that the idea of the event in a strong sense is now typically suppressed by ideologies and practices that render fundamental change almost unthinkable. Impulses that cannot otherwise be contained are channeled economically into thirst for the latest device, fascination with the newest sport or singing sensation. They can be diverted "politically" into electoral charades that, whatever their rhetoric, finally offer nothing more than continued *stasis*, in both the ancient sense of irresolvable conflict and the modern one of an immovable *status quo* (or gridlock). This last is typically the effect of a struggle in which both sides have vested interests in keeping things within fairly narrow bounds, as in parliamentary democracies' circular dance of center-right and center-left parties.

I propose that a more productive reading of Nietzsche will attend both to his thought of futurity and his call to be true to the earth, as in his lapidary but still little-understood declaration that only from his work on will there be a great politics on and of the earth. As I've hinted, talk about the futurity of the earth involves what may seem like unnecessary excursions into questions of linguistics and etymology. Some would say that the first issues on the agenda concerning a politics of the earth should be climate change, energy needs, globalization, and geopolitical conflicts still inflected and infected by religious and ethnic hostilities. While Nietzsche does have things to say that could help to open up our thinking about such questions, as in his notion of the human-earth as a garden, I propose that we follow the "old philologist's" advice to begin by attending carefully to his words and discourse. So I forewarn the reader that a good deal of this book is concerned with explicating terms crucial for Nietzsche's multilayered thought concerning the *Sinn der Erde*. That phrase itself, one Zarathustra introduces in his first public discourse, requires to be heard with care. It is usually translated as "the meaning of the earth." Yet *Sinn*, as Günter Figal reminds us, also signifies direction.[4] Where is the earth going? In what direction will you deploy your energies for earth's sake? To be loyal to the earth, to give it your *Treue* (or troth), means to accept discipline, to be ready to sacrifice. And how should we understand Nietzsche's concept of the earth? I will argue that above and beyond what we might call its phenomenological sense as our immanent lifeworld (the limit of most scholarly readings), the earth in Nietzsche's writings has a political sense as the counterconcept to what Hegel and Hegelianizing philosophers call the world. Hegel's concept of world, we will see, is a unitary notion. It cannot be decoupled from those of the state, world history, and God. It is ultimately a concept of political theology, which finally provoked Nietzsche to articulate a philosophy of the Antichrist. When Nietzsche speaks of the earth (sometimes more specifically of the *Menschen-Erde*), he is at least implicitly formulating a political atheology, an understanding of the sphere or territory of human habitation; Nietzsche's war for the sake of the earth must involve an attack, parody, and inversion of political theology. The earth in this perspective is radically plural. It is neither intrinsically defined by the nation-state (like Hegel's world), nor, as in the *Weltprozess* of Eduard von Hartmann (the largely forgotten target of Nietzsche's *Unmodern Observation* on history) the site of an inexorable teleology. Such a contrast of earth and world is very close to Deleuze and Guattari's methodological protocol of subordinating history to geography.

This book can be read then as a series of philological commentaries, taking philology in Nietzsche's sense of a critical discrimination of meanings and texts informed by a genealogy of power. The most schematic form of these commentaries revolves around five contrasting pairs of terms, including world and earth. The others, to be explored in some depth, are: states and nomads, masses and multitude, *kairos* and *chronos*, Christ and Antichrist. In many cases even some of Nietzsche's most astute readers have neglected or even seriously mistranslated some of these. For example, Nietzsche says emphatically in a crucial passage of *Beyond Good and Evil* that "this is the century of the *multitude* (*Menge*)" (*BGE* 256), but even respectable translators render this as "masses," although elsewhere Nietzsche repeatedly makes a clear distinction between the two terms with respect to the masses' homogeneity and the multitude's diversity.[5] Other scholars blunt the force of Nietzsche's deliberately outrageous invocation of the ominous Christian figure of the Antichrist, with its accumulated legends and its crucial role in Christian political theology, by rewriting the term as "anti-Christian." The latter is a possible meaning, and is indeed ingredient in the personification of Christ's opposite and ultimate enemy, but it is not what Nietzsche intends, I'll argue, when he speaks of a "philosophy of the Antichrist" in the aphorism that concludes his examination of "Peoples and Fatherlands." The Antichrist is the lord of the earth, an earth that persists and eternally recurs in Nietzsche's atheology, rather than passing away, as in the Biblical text (Apocalypse or Revelation) he parodies.

Reading Nietzsche philologically should also involve reading him historically, that is, triangulating his thought between his time and ours. Nietzsche was responding both explicitly and implicitly to themes and problems relevant to both: nationalism, the consolidation of state power, the theory and practice we've come to call globalization, the dispersion and nomadic movements of peoples, the threat of enormous unpaid and unpayable debt, the emergence of mass media and entertainments, the continuing power of older religious hatred and the looming possibility of nihilism, which he sometimes describes as European Buddhism.[6] As cultural physician, Nietzsche diagnoses his time by taking his scalpel to lay bare and dissect its (presumed) virtues. While we rightly suspect some of his radical suggestions for a cure, we still have work to do to decipher the language of his analysis and prescriptions, and the problems do not all arise from Dr. Nietzsche's notoriously difficult handwriting on the Rx forms (or to put it more prosaically, his finely tuned style).

Consider, for example, the contours of Nietzsche's conception of the earth—sometimes called with emphasis the human-earth. These emerge more clearly when we see it developing as a running critique of the Hegelian idea of world-history. The nineteenth century was the era of world-history, with philosophy morphing into journalism as both professors and popular writers competed to provide the most up-to-date, modern accounts (*zeitmässig* is Nietzsche's term) of the meaning of history. Nietzsche himself, in his later preface to *The Birth of Tragedy*, confessed that he too had given in to such world-historical temptations when he foresaw a Wagnerian cultural renaissance. Whether the meaning of history is thought to lie in democracy, socialism, or technological "progress," it seems important to bolster the sense of inevitability with a persuasive and seductive metanarrative. It is not only Marxist socialism that has followed this path. The brand of American exceptionalism that from Woodrow Wilson on heralds the United States as the avant-garde of a globally irresistible democratic freedom has a similar (if non-dialectical) deep structure, as do the many technocratic fantasies of total mastery of nature.

"Earth," I want to suggest, is not so much the *telos* of Nietzsche's own metanarrative as the signature of his alternative to metanarrative, a genre he sees as essentially Christian (and thus vulnerable to being undermined through his inversion of the Antichrist *topos*). In writings of his last productive years, Nietzsche regularly associates earth and the political. "Only after me will there be *great politics* on the earth," Nietzsche writes in *Ecce Homo*, as he explains, "Why I Am a Destiny (1)."[7] Much thought has been expended in attempts to make sense of the content of "great politics." Is it a politics of race, nations, class, or ideology? Is it a politics aiming at helping Europe to become one, perhaps under the sway of a new ruling caste? Where does great politics lie on a spectrum that includes tyranny, mass democracy, and anarchism? Are these the right questions to be asking? As Bruno Bosteels observes, there is a great disproportion between Nietzsche's few brief references to great politics and the many lengthy commentaries on this theme.[8] Since many of Nietzsche's dramatic claims of this sort were made in the last year before his mental collapse, in writings marked by a hyperbolic sense of his self-ascribed importance, it is tempting to discount them as symptoms of the coming personal catastrophe.

We might read Nietzsche's rhetoric of political catastrophe, involving new kinds of wars and the total trembling of old orders, likened to earthquakes, as transcriptions of his individual mania onto a larger canvas. I resist such premature psycho-biographical reductionism, while noting that current fears of earth's ruination through climate change, pandemic, overpopulation, and new forms of war have overtones of the apocalyptic motif that Nietzsche evokes in deploying the Antichrist theme.

For now I propose to minimize two forms of speculation that either ascribe a specific political program to Nietzsche or that focus on his personal medical and psychological condition. Rather, I return to what may seem like a narrow philological observation. What is typically neglected in Nietzsche's cluster of assertions about great politics are the frequent references to the ultimate subject of such politics, the earth. In the *Ecce Homo* passage quoted above, it is the *earth* that will know great politics for the first time, and in the section leading up to that declaration he says that his name will be associated with a crisis "like no other on earth." As if to emphasize the earthiness at stake here, he says, "We shall have convulsions, an earthquake spasm, a transposition of valley and mountain such as has never been dreamed of." So what, goes the simple response, Nietzsche obviously means to emphasize the global scope of this politics, its implications for the entire world. Yet in fact Nietzsche is often suspicious of talk about the world (*Welt*), especially when it is associated with world-history (*Weltgeschichte*) or the supposed world-process (*Weltprozess*), the mushy concept of Eduard von Hartmann that he subjects to withering sarcasm in the second *Unmodern Observation*. Very frequently, and increasingly in his later writings, Nietzsche chooses to speak of earth rather than world when the topic has political implications. And of course the earth, sometimes named as the human-earth (*Menschen-Erde*), is the initial concept of Zarathustra's first public address, when he implores the crowd to swear their fidelity (*Treue*) to the earth. World is tied up with the idea of world- history, involving the notion of a teleological development realizing an inevitable end. For Hegel himself, the world is composed of a system of states, ordered in an evolving system. However much this world-history is a "slaughter bench," as Hegel famously said, it is also true for him that world-history is "God's march through the world." Less well known is that Hegel says that those people without a true state are also lacking a world.

Nietzsche's idea of earth, I'll be arguing, is formed as a political alternative to the Hegelian conception of the world that was taken for granted by so many of his contemporaries, although often in diluted form. Many of Nietzsche's readers emphasize what might be called the phenomenological aspect of his idea of the earth as a dimension of radical immanence and bodily life; as he says in *Zarathustra*, earth must be defended from those who preach imaginary worlds behind the scenes, beyond all possible experience. In the book he called the greatest gift ever made to humanity, this is evident at the most superficial level. As a phantasmagoric landscape poem, *Zarathustra* draws on several rich traditions of writing about the earth, showing its hero climbing up and down mountains, responding to changing seasons, climates, times of day from dawn to dusk and noon to midnight, wandering through towns, sailing the seas, exploring islands, possibly descending into a volcano, and celebrating the thought of eternal recurrence with a song ("The Seven Seals") that finds him identifying with a bird's freedom and flight in an unbounded cosmos. Of course this earth of experiential immanence provides the indispensable condition of *grosse Politik* as well as its theater. For Nietzsche, the human-earth is an inexhaustible source of nourishment and stimulation on all levels, from bodily sustenance to adventurous challenges. It is the resource of all resources and so should become the subject of great politics. The earth then is the prime object of the political economy implied by the concept of an *Umwertung aller Werte*. The earth is the ultimate focus of all orientation. Even if the madman is right in saying that without God, and with the post-Copernican decentering of the earth, we are adrift in measureless space, we are still adrift *on the earth*. As Husserl says in a late essay, the earth by which we orient ourselves does not move. Even if we should leave the earth, we will still orient ourselves in relation to it.[9]

The earth is not simply that which supports us and by which we orient ourselves. It is also that with which we ceaselessly interact. We are constantly moving on its surface, cultivating it, tunneling and mining it, and extracting its resources. Nietzsche has Zarathustra describe humans as a skin disease on the earth, but he also imagines the earth transformed into a gigantic health resort and tree of life (*Z* II.19; *WS* 188–89). His *Menschen-Erde* could be a translation of the recently named geological era of the anthropocene, roughly the time since the last Ice Age, during which the earth begins to undergo transformation through human activity such as the agricultural and urban revolutions.[10] Surely this conception of the anthropocene was anticipated and known to Nietzsche through the famous choral ode of Sophocles's *Antigone*.[11] The human race is uncanny, terrible, and wonderful (*deinon*) in traveling the seas, plowing and transforming the earth, capturing and taming birds, fish, and beasts, and building cities. The chorus's attitude is not only triumphal, for humans engage in these enormous seizures, dislocations, and alterations for both good and ill. Nietzsche's directive to be loyal to the earth and vigilantly mindful of its direction must be read in this tragic perspective.

When he writes in his last lucid months of a great politics of the earth, involving a trembling and shaking so far unknown, Nietzsche restates and condenses a major theme of his thought, one he has been developing for years, into a handy formula. A great politics of the earth certainly contrasts with a small politics, one focused on the petty and confused doings of states, national rivalries, and crude demagoguery such as the anti-Semitism he came to despise. Nietzsche can amuse himself with such small politics by reading the "news of the [German] *Reich*" while lounging in Venice's Piazza di San Marco (*GM* III.8) or dismiss it as mere "contemporary chatter (*Zeitgeschwätz*)" (*AC*, P) as he declares himself the Antichrist in what turned out to be his final politico-*a*theological treatise.

Nietzsche's challenge to his age involves a rejection of world-history and its claim to have understood the structure of time. Being true to the earth necessarily involves abandoning the conception of time that subordinates earthly life to a metanarrative in which that life is merely the frame for the eventual manifestation of the Idea or the Christian end of days (an identification made explicitly by Hegel). Now if the most significant name for the world's time is world-history, what is the time of the earth? This, I think, is the question motivating the diptych of *Zarathustra* and *Beyond Good and Evil.* The first, a "yes-saying" prophetic manifesto for the earth affirms earth's eternal recurrence, envisions its transformation into a garden, and introduces the idea of a great event, while the second "no-saying" critical analysis insists on the ineluctability of futurity and offers advice on the vigilance necessary to seize the *kairos*.

World-history, by the way, is a pleonasm and tautology, something that becomes clear and explicit in Hegel. A world, so far as it consists in peoples organized into states, sharing a culture, practicing and professing a religion, can and must have a history until that history is fulfilled and completed. Conversely, the only thing that can have a history (as opposed to a logical development) is a world composed of such politico-theological-cultural units. World-history proceeds through a series of contradictions that must be resolved or accommodated, typically by the production of new arrangements with their own tensions. For Hegel, the Greek contradiction is set out dramatically in the tragic confrontation of the state's public law and the family's private religious law; he famously reads this structure in Sophocles's *Antigone*. That conflict yields to the Roman state, with its emphasis on private persons and property on the one hand and public rule on the other, finally expressing itself fully in the contradiction of the many individuals and the supreme imperial embodiment of authority. This sets the stage for Christianity, in which the divine emperor is seen as merely a stand-in for God, and his connection with the many equal individuals is mediated by his son. Nietzsche highlighted this step in Christianity in his 1886 preface to *Dawn* (five years after its initial publication). The book's title can be read as the antithesis of the Hegelian twilight at the end of history when the owl of Minerva spreads its wings. Following its epigraph from the *Rig Veda*—"there are so many dawns that have not yet broken"—it would mark the possibility of new beginnings and futurity in opposition to any thought that history is complete and reason has pronounced its last word. Nietzsche (perhaps abusively) traces "German" dialectical logic to the version of Christianity whose principle is "*credo* quia *absurdum est*"—"I believe *because* it is absurd"—announced by Tertullian and others.[12] This saying now makes us think of Kierkegaard, perhaps an even more ferociously anti-Hegelian thinker than Nietzsche; as an advocate of this profession of faith, he sees it as wholly contrary to Hegelian reason. Kierkegaard sought to undermine Hegel's claim that his logic and system were thoroughly Christian. Yet here Nietzsche first ascribes an extreme antirational fideism to (authentic) Christianity and then identifies it with Hegel's logic, as if he were collapsing a merely apparent distinction between Hegel and Kierkegaard.[13]

From Nietzsche's perspective, Hegel did not rationalize Christianity but disguised its true antirational core—celebrated by Tertullian, Luther, and Kierkegaard—as the highest reason. The very rhetoric of his critique, like Hegel's, speaks of "German soul" and "German spirit." Nietzsche out-Hegels Hegel, we might say, in identifying his logic, in a way that is ultimately geophilosophical rather than world-historical, as "German logic":

> Nothing, of course, has ever made a deeper impression on the German soul, nothing has "tempted" it more than this most dangerous of all conclusions, which to every true Mediterranean, is a sin against the spirit: *credo* quia *absurdum est*—with this conclusion German logic makes its first appearance in the history of Christian dogma: but even today still, a millennium later, we Germans of today, late Germans in every respect, catch the scent of—something like truth, like the possibility of truth behind the famous real-dialectical axiom with which Hegel in his day procured for the German spirit [*deutschen Geist*] a victory over Europe—"Contradiction rules the world: all things contradict themselves"—we are, through and through, even in our logic, pessimists (*D* P3).

In 1886 then, Nietzsche would have us read his 1881 text in terms of the difference between the Mediterranean and German spirits. Like Hegel, Nietzsche names "spirits" that are strongly associated with geographical regions, nations, and, in some eyes, races (by 1886 Nietzsche had long been disabused of any metaphysically suspect idea of race, although he entertained some speculative notions of individual lineage that might now be expressed in terms of DNA transmission).[14]

Rather than agreeing with Hegel that his dialectic allows the implicit rationality of Christianity to become explicit, Nietzsche asserts that his predecessor follows an antirational thematic already sounded by Tertullian, Luther, and others. Despite having written about the primordial contradiction at the heart of things in his first book and in his fervent praise of Heraclitus's teachings—which Hegel claims to have incorporated into his own logic—Nietzsche here seems to turn against a related form of thought. We might begin to articulate the differences by distinguishing between contradictions or absurdities as such (Nietzsche frequently acknowledges being as irreducible chaos, e.g., *WP* 1067) and a logic of contradiction that issues in a system. If Nietzsche is thinking of Tertullian, to whom the formulation of the *credo* is attributed, it is noteworthy that this early Church Father is also said to be the first Christian theologian to speak of the Trinity. It is not just any or all contradictions that issue from his embrace of a fundamental absurdity, but the outlines of a systematic theology. Indeed, Hegel repeatedly tells us that the Trinitarian logic of Father, Son, and Holy Spirit informs all of his own system.

In Hegel's world-historical narrative, the German spirit takes up Christianity and in doing so transforms both itself and the religion it is assimilating in the direction of universality. In this story (considered later in more detail), a religion that emerged from the contradictions of the Roman world is absorbed by Germanic tribes, who have the vitality and flexibility to give it a coherent shape as a world religion. For both Hegel and Nietzsche the distinction between the Mediterranean (or Hegel's Greek and Roman worlds) and the Germanic spirits is crucial. The spirits they analyze are both associated with geographical regions. While most of the national spirits that Hegel examines simply stay in their places (or, like Arab-Islamic culture supposedly dissipate themselves because they lack a territorial foundation), it is uniquely the German spirit that becomes truly universal. Nietzsche leads German logic back to its cultural and geographical limits, contrasting it with Mediterranean thought, applying to Hegel something like the system of cultural classifications that Hegel applies to others.[15]

In the 1881 text of *Dawn*, Hegelian world-history is an early target. Moreover, the very form of this aphoristic text contests not only Hegelian systematicity but (as the 1886 Preface indicates) it should be read slowly (*lento*) and philologically, thus enacting a specific temporality that does not rush to conclusions, leaving the future open. *Dawn* is informed by ethnology and anthropology; Nietzsche marshals anecdotes from these emerging human sciences to show that what is called world-history is only a very narrow tranche of the larger history through which human beings have shaped themselves and their cultures. One of the earliest and longest sections of the book makes this point in the course of arguing that we fail in our self-knowledge as a species if we are oblivious of the many millennia during which cruelty to self and others formed the human beings who could only then become the putative subjects of world-history. World-history then fails to understand some very important things about time, beginning with its extreme compression of meaningful human development into a small fraction of a more general history:

> Every one of the smallest steps along the path of free thought or in a life shaped personally has been achieved, from time immemorial, at the price of spiritual and physical tortures . . . something we certainly forget when speaking, as usual, about "world history," about this ridiculously tiny slice of human existence; and even in this so-called world history, which is basically just noise around the latest news flashes, there is actually no theme more important than the ancient tragedy of the martyrs *who wanted to stir up the quagmire* (*Sumpf*) (*D* 18).

What little bit of reason and freedom we humans now possess, Nietzsche continues, are legitimate sources of pride. They are not built into history as its leading motives, but are fortunate achievements of a few spiritual pioneers who provoked change by seeking or accepting unusual forms of suffering and *askesis*.[16] Reason and freedom, of course, are the leading themes of fetishized world-history, sounded as a constant drumbeat in self-congratulatory versions like Hegel's. However, a tragic dimension, involving pride and the failure of self-knowledge, has been repressed and not fully *aufgehoben* in Hegel's ultimately comic story. The happy ending of that narrative, with its successful drama of recognition, makes it all too easy for us to see how limited the story is:

> It is, however, precisely this pride that makes it virtually impossible today for us to have a feeling for those vast expanses of time that comprise the "morality of mores [*Sittlichkeit der Sitte*]" and that precede "world-history" as the *actual and decisive main history* [*Hauptgeschichte*] *that has determined the character of humankind* (*D* 18).

That story of world-history, the ground of the world-spirit, is the history of states, and a true state, on Hegel's account requires a religion and a monarchical order. Deleuze and Guattari credit Nietzsche as the founder of geophilosophy, an approach that avoids teleology and statism, emphasizing rather a plurality of forms of human earthly habitation, including nomadic and other nonstate groupings.[17] Eventually, as in Zarathustra's talk on "The New Idol," he will say in effect that the earth's future is too important to be left to the state. Yet Nietzsche came to see that he had not always been free from Hegelian thinking. Ironically, the first of his *Unmodern Observations* was a demolition job on D. F. Strauss, in which he argued that Strauss was self-deluded in thinking he had freed himself from Hegel. Yet in 1886 Nietzsche confessed that *The Birth of Tragedy* was still far too Hegelian for his later taste. He sees his own trajectory as a progressive catharsis of the Hegelian infection of the nineteenth century.

As we will see, Wagner becomes a signature of Nietzsche's struggle with Hegel, including the Hegel that both had unconsciously absorbed. Through *UO* IV, Wagner is explicitly seen as the anti-Hegel, yet Nietzsche uses suspiciously Hegelian arguments to establish Wagner's cultural leadership. In the later work, Wagner and Hegel are identified (*CW* 10). This ambivalence about the Hegel-Wagner relation accounts for the abrupt termination of the series of thirteen *Unmodern Observations* Nietzsche had projected after the fourth, *Richard Wagner in Bayreuth*. Yet the anti-Hegelian program of *Richard Wagner in Bayreuth*—whose markers are the event and a nascent geophilosophy—returns in *Thus Spoke Zarathustra*.

While the critique of world-history is already there in the 1881 text of *Dawn*, the 1886 preface shows that it has become more explicitly geophilosophical (borrowing a term from Deleuze and Guattari, who call Nietzsche the inventor of geophilosophy). One terminological mark of this intensification is the increased use of the term earth (*Erde*) to name the most fundamental arena of human activity. Zarathustra's initial and primary message is the call to be true to the earth. He voices apprehension for the fate of the *Menschen-Erde*, the human-earth, which can manifest itself either as a hell or a garden, possibilities that are explicitly articulated in "The Convalescent," a chapter generally regarded as one of the most significant in Nietzsche's "greatest gift." Many have tended to opt for either a phenomenological or hermeneutic approach to understanding the earth and the elemental. At least they have heard the words, although they have often failed to articulate them in phrases involving "great politics" and the like. Returning to the phrase with which we began, "only beginning with me is there *great politics* on earth," it is remarkable that recent discussions, especially Anglophone ones, typically pass over the *subject* of great politics, the earth. What does it mean to say that the *earth* will have great politics? Is it simply that from Nietzsche's time on, and because of him, in his name, a new great politics will arise, a politics that differs in important respects from a small politics? Certainly the statement speaks of a new, coming politics, one that will not merely occur or be practiced but will be recognized as such on a large, perhaps the largest, scale. Does the use of the word *earth* simply indicate generality? If Nietzsche has a novel conception of *grosse Politik*, might he also have a conception of earth that needs to be explored?

Certainly the meaning of "*grosse Politik*" has been controversial. Let us set aside the polemics that immediately identify Nietzsche with the politics of Wilhelm II or Hitler, while not ignoring the question whether "great politics" could be conducive to tyranny and unlimited conquest. There is then much discussion about just what is "great" about "great politics." Minimally, this great politics contrasts with the petty politics of European nation-states. While Nietzsche's observation that "Europe wants to become one" has been read as expressing a related thought, those who explore the implications of Nietzsche's thought for a "new Europe" tend to ignore the question of whether and how this is a politics of the *earth*. Although it was "*great politics*" that Nietzsche emphasized in his text, it is the *earth* that is its *site* and knows it *only with him*. If the earth is left out of account, we would be left to focus either on the "content" of great politics or on the event of its enunciation, the event named Nietzsche. Small politics is a Westphalian politics of states; great politics involves "wars the like of which have never been seen on earth." Are these wars for ideas as Nietzsche sometimes suggests, perhaps the final battle of the Crucified and Dionysus, or Christ and Antichrist? Should we think that Nietzsche is promoting a new, larger-scale, earth-wide version of tyranny, aristocracy, or democracy? Is his central political notion that of a new ruling caste for Europe (or the earth)? These are valuable and important questions that are subject to a long and still vigorous discussion. I neither dismiss this discussion nor presume to provide new answers. Rather, I want to suggest that without articulating the role that the earth plays in *grosse Politik* we should be hesitant in pinning anything like these traditional labels on a figure who imagined himself as an uncanny and monstrous destiny.

Nietzsche began to use the expression "great politics" in *Human, All-Too-Human* (1878). He was responding to the journalistic discussion of Bismarck's policies, which was given this label in the press. As Peter Bergmann documents in his indispensable study of "the last antipolitical German," Nietzsche is contemptuous of such claims to greatness, viewing Bismarck and his ilk as simply continuing the practices of nineteenth-century nationalism.[18] In a few years he will write an aphorism in *Dawn* with the title "Great politics" that comments on how the masses find emotional exaltation and a sense of purpose as politicians marshal and provoke nationalist feelings. Nietzsche comes to adopt the term as his own to designate a more authentic form of "*grosse Politik*" in an 1884 letter to Overbeck.[19] He goes public with this usage in *BGE*, where he says "the time for petty politics is over: the next century will bring the struggle for dominion of the earth (*Erd-Herrschaft*)–the *compulsion* to great politics" (*BGE* 208). Great politics in Nietzsche's writings from at least 1884 on means a great politics of the earth.

I will not provide a detailed analysis of the enormous range of what has been said about Nietzsche's "great politics" since his claim to mark its initiation in 1888 (in what he calls "the old time scheme"). Consider some representative treatments that tend to downplay the earth's role in Nietzsche's "great politics." Vanessa Lemm argues that Nietzsche's great politics should be understood as a great politics of the event, where the event is understood as the production or emergence of great humans. However, she also argues that "Nietzsche's great politics of the event qua cultivation of the great human being does not mean an elitist politics of domination."[20] While Lemm refers briefly to *Zarathustra*'s chapter "On Great Events," she does not specifically address the question of earth's status there, a status emphasized in the geographical and geological framing of the chapter and Zarathustra's emphatic declaration, meant to silence the howling of petty politicians and self-inflated revolutionaries, that "the heart of the earth is of gold."

Alain Badiou has written an essay that puts a Nietzschean claim in the interrogative mode, asking *Casser en deux l'histoire du monde?*[21] Nietzsche purports to accomplish this split, in Badiou's analysis, not merely by developing concepts but through a philosophical act that is itself without meaning, and so can only be named; it is the very name "Nietzsche" that designates the act. What strikes readers of the late texts as a sign of madness is, for Badiou, a lucid way of announcing an insight into the character of the act as such. He proposes that Nietzsche's "Why I Am a Destiny" can be explicated as "I am a destiny because, by chance, the proper name 'Nietzsche' comes to link its opacity to a break without program or concept."[22] If philosophy after the French Revolution presents itself either as a restoration of order (e.g., Hegel, Comte) or as a radicalization of the revolution (Marx), the Nietzschean act is a rival of the revolutionary act; it is "arch-political, in that it proposes itself to revolutionize all of humanity on a more radical level than all of the calculations of politics." Badiou reads *Zarathustra*'s chapter "On Great Events" as a confrontation with the revolutionary political event in the form of the fire-dog.[23] In his readings of Nietzsche's remark about splitting the history of mankind in two, and his claim that only with him will earth know great politics, Badiou shows no interest in the roles that the crucial terms *world* and *earth* play in these texts. The dialogue "On Great Events" is informed by its framing on a volcanic island; Zarathustra's final emphatic words to the fire-dog are "*the heart of the earth is of gold.*" However indeterminate the "great event" may be—what else could it be, given Nietzsche's insistence on the radical openness of the future?—it is not simply Herr Nietzsche's event, but an event of the earth, contrasting with those of the world, in the limited sense of that term.

Rather than asking about the degree of Nietzsche's authoritarian or tyrannical tendencies, or the question of who would assume rule or hegemony over the earth, I propose to explore what he means by *earth*; in other words, what is the theme or subject of his political thought? It should not be surprising to hear that earth is a fundamental political concept for Nietzsche; after all, his hero Zarathustra begins his descent (or reentrance) into human communities by seizing the opportunity to speak in the public forum of the marketplace, challenging the assembled multitude to dedicate themselves to the direction or meaning of the earth (*Sinn der Erde*). Yet surprisingly little attention has been given to clarifying Nietzsche's concept of the earth. While "earth" is often taken in a general way to refer to embodied life, to *this* world rather than to an imaginary and disastrous *other* world, I propose that the term and concept also have a significant political dimension—a geophilosophical dimension—which is closely related to the radical immanence so central to Nietzsche's thought. I shall argue that he often and pointedly replaces the very term "world (*Welt*)" with "earth (*Erde*)" because "world" is tied too closely to ideas of unity, eternity, and transcendence.

"World" is a concept with theological affiliations, as Nietzsche indicates in *Beyond Good and Evil*: "Around a hero everything becomes a tragedy, around a demigod everything becomes a satyr play; and around God everything becomes—what do you think? perhaps the 'world?'" (*BGE* 150). This can be amplified when we recall Nietzsche's declaration that he is afraid we haven't gotten rid of God yet, because we still have faith in grammar, or his thesis that with the disappearance of the "true world" the apparent one disappears as well, and in his speaking of the lingering shadow of God (T "World," *GS* 108). The trinity of God, human, and world is a common, traditional complex of philosophemes. Perhaps one of the late-arriving insights that follow in the slow mourning process that accompanies God's death has to do with the disappearance of that which we call world. Like all metaphysical and theological concepts, *world* has a political import, one evident to Nietzsche in Hegel and those he considered Hegelians (e. g., D. F. Strauss and Eduard von Hartmann, targets of his first two *Unmodern Observations*); in *The Birth of Tragedy* he speaks contemptuously of "so-called world-history" (*BT* 15) and in *The Advantage and Disadvantage of History for Life* he ridicules the fashionable notion of the *Weltprozess* (an anticipation of the notion of globalization) and exclaims "world, world, world!" in high dudgeon (*UO* II.9). When Nietzsche comes to write of "great events," they are not tied exclusively to the state and world-history, as they are for the Hegelians, but (as "*On Great Events*" makes clear) events of the earth. If for Hegel "the state is the march (*Gang*) of God through the world," for Nietzsche the earth is a human-earth of mobile multitudes that can prepare a way for the overhuman. In order to grasp Nietzsche's "great politics" of the earth more perspicuously, it is useful to see how both its rhetoric and substance constitute a response to the theologico-political treatise that is Hegel's *Philosophy of World History* and to those Nietzsche saw as Hegelian epigones.[24]

Since Nietzsche claimed that *Thus Spoke Zarathustra* was his most important work, we must keep in mind the intensity of Zarathustra's striking invocations of the earth there. He calls on his listeners to sacrifice themselves for the *Sinn der Erde.* While this phrase is typically translated as "meaning" or "sense," it could also be rendered as "direction." Where is the earth going? Where do we want it to go? Zarathustra requires his disciples (*Jünger*) to give their loyalty (*Treue*) to the earth, addresses the condition of the human-earth (*Menschen-Erde*), and encourages his listeners to think with "an earthly head that creates a direction for the earth! (*einen Erden-Kopf, der der Erde Sinn schafft!*)" (*Z* P 3; I.22; III.13.2; I.3). The earth—not the world—must be rescued from the threatened domination of the last human: "For the earth has now become small, and upon it hops the last human, who makes everything small" (*Z* P 5). After *Zarathustra*, Nietzsche's later works typically refer to a project of evaluating moralities, religions, and cultures as ways of being "on the earth": I will show that this is typically more than a conventional phrase. Nietzsche evaluates the forms of human life by asking whether they are worthy of loyalty and sacrifice, whether they have a worthy future direction, and whether they can ward off the reduction of spirit and adventure that rationally securitized technopolitics is now threatening.

Many critical engagements with Nietzsche's idea of earth tend to ignore or minimize the political, geographical, and geological relevance of the concept. One approach sees earth as designating the immanent, bodily, or this-side, as opposed to the imaginary beyond of religious and transcendental traditions; while not inaccurate, this characterization remains rather indeterminate.[25] A phenomenological interpretation emphasizes Nietzsche's poetics and metaphorics of the earth, sometimes enriched by recalling his experience as traveler, walker, and poet receptive to the beautiful, sublime, and picturesque in natural and artificial landscapes.[26] Such readings include Gaston Bachelard's celebration of Nietzsche's virtual flight (air as an earthly element) and Luce Irigaray's disappointed love letter, lamenting his avoidance of the feminine, maternal sea.[27] Some hermeneuts focus on Nietzsche's adaptation of poetic and philosophical topoi from early Greek thinkers and poets, especially Empedocles, for whom Gaia retained features of the divine.[28] Inspired by Nietzsche's reading of Hölderlin, and Heidegger's reading of both, this approach tends to stop short of articulating the way in which, thinking with his *Erden-Kopf,* Nietzsche conceives the *Sinn der Erde* against the background of Hegelian philosophy of history, the nineteenth-century state, nationalism and its complicity with romanticism, and the tradition of political theology. Nor does it take note of his recognition of new paths developing in human geography that highlight human mobility: nomadism, migrations, and wanderings of peoples.[29] These classicizing and romanticizing readings also obscure the ways in which Nietzsche offers a very specifically earth-centered reversal of the Biblical themes from the innocent garden to the terror of apocalypse and Antichrist.

The political dimension of Nietzsche's concept of earth comes more clearly into focus if we ask what ideas, concepts, and theories it opposes and which it means to undermine. Otherwise said, I propose to hear Nietzsche when he writes of the earth not only as designating the *Diesseits* rather than the *Jenseits*, or expressing a poetic sense of the textures of immanent life, but as also (at least frequently) referring to a fundamental dimension of human political life, conceived differently than the world of the nineteenth-century's world-history. This involves understanding Nietzsche as engaging critically in a contemporary set of discussions about issues that have more recently been named by terms like the end of history, globalization, states of exception, the multitude, and political theology. While this is the language of philosophers and political theorists like Alexander Kojève, Francis Fukuyama, Giorgio Agamben, and Carl Schmitt, not the vocabulary of the 1870s and 80s, I will be arguing that the concepts, if not the names, are crucial to understanding Nietzsche's political thought. Further, I will show that Nietzsche's conception of earth's direction depends upon a complex set of ideas about events, time, and temporality that constitutes an alternative to the way in which the theorists of world-history construe these ideas.

The Hegel whom Nietzsche confronts will strike some readers as a caricature, based on a selective reading of incomplete and questionable versions of his lectures. But then perhaps much of the nineteenth century could be called a caricature of Hegel—and should not he and Nietzsche bear equal responsibility for those who caricature their thought? While more recent scholarship has given us a more subtle Hegel—actually, a choice among several versions of a more subtle Hegel—Nietzsche's Hegel is firmly based in the text of *The Philosophy of World History* that was available to him. The popular Hegelians of Nietzsche's day—e.g., Strauss and Hartmann—reinforced the caricature, if such it is, and made it a forceful presence in Nietzsche's Europe—especially Germany. Finally, I believe that it will be easier to see what Nietzsche was arguing against, when we recognize that much recent scholarship has been overly zealous in its attempt to provide a Hegel who would be congenial to a democratic, pluralistic era, even to the point of producing somewhat misleading translations of key titles and passages. For example, the only English translation (Sibree's outdated mid-nineteenth-century version) that purports to render the whole of Hegel's lectures on *Weltgeschichte* drops the "world" in the title, rendering it simply as the *Philosophy of History*, and even recent partial translations of the introduction to these lectures follow Sibree's lead.[30]

Let me clarify a few questions of method or reading protocol. Some readers may suspect that I exaggerate the distinction between earth and world, or am introducing a distinction Nietzsche never made. Surely Nietzsche or his surrogates (Zarathustra and the many voices deployed in other texts) often praise "this world" or simply "the world" in opposition to imaginary worlds behind the scenes. Admitting the charge of exaggeration, I claim that the distinction is one that is implicit from the beginning and becomes increasingly more prominent in Nietzsche's work. The exaggeration makes explicit a constellation of related themes in his thought, which form the subject of the following chapters. While Nietzsche is almost always scornful of world-history, he consistently follows the spirit of Zarathustra's injunction to be true to the earth, insofar as he tends to give this name to the broadest sphere of human action and history. In the *Genealogy of Morals*, regarded now by many as Nietzsche's most polished and coherent book, the rhetoric of earth is pervasive. Earth is clearly the site and focus of the analysis, whether he is describing it as the "ascetic star" (*GM* II.11) or lamenting that so far "the *will* for the human and the earth has been lacking" (*GM* III.28). Philosophy is said to appear on the earth, not in the world (*GM* III: 7, 9); religious history is summarized by saying that "the earth has been a madhouse for too long!" (*GM* II.22); and the author looks forward to a future "that gives back to the earth its goal and to the human its hope" (*GM* II.24). While "world" also occurs a number of times in the *Genealogy*, these usages are more general, as in references to "the modern world" or something "coming into the world"; when Nietzsche uses the term in a more determinate way, he quotes his own *Gay Science* to speak of the difference between "this world" and the "other world" (*GM* III.24). Nietzsche performs an exemplary deconstruction of this true world/apparent world dichotomy in *Twilight of the Idols*, in his compressed narrative of "How the True World Became a Fable," which demonstrates how the two ideas are structurally and metaphysically interdependent and so can both be eliminated. There is in Nietzsche no corresponding pair of an apparent and true earth. To find such an opposition we must go to Platonic myth. Paul Loeb has suggested that the concluding aphorisms of *GS* IV, which introduce the Zarathustra theme, show that Zarathustra is in many ways a response to the Platonic Socrates of the *Phaedo*, who serenely welcomes death as a release from life. We could add that part of Socrates's swan song is his story about the true earth that is vastly superior to the limited swampy part we ordinarily know.[31] Yet for Zarathustra and Nietzsche there is no such distinction. However much the earth has been degraded and corrupted, it retains possibilities of renewal and growth, or as Zarathustra says to the world-crazed fire-dog, "the heart of the earth is of gold" (*Z* II.19).

This book aims to sharpen thinking about both contemporary planetary crises and Nietzsche's view of the political by explicating his injunction to give the utmost priority to the question of the direction (*Sinn*) of the earth. Nietzsche, I argue, saw such developments as globalization, new forms of war, transformations in the institution of the state, the emergence of nomadic and hybrid human types, and the increasing realization of the fragility and possibilities of the human environment as interlinked phenomena. In making this case, and in clarifying his analysis, I will draw on his often-overlooked relation to the nineteenth-century context in which he wrote, while showing that his concerns are worthy of thought today. I triangulate, reading Nietzsche in the context of both his time and ours. Nietzsche is one of the very few major philosophers to have taken on questions with global scope when it first became possible to do so (given nineteenth-century developments in industry, transportation, and communication). He has important things to say about topics clearly on the contemporary agenda: the explicit and implicit presuppositions and consequences of globalizing "end-of-history" narratives; the character of a livable and desirable earth (a "human-earth"); the geopolitical categories of "peoples and fatherlands," states and nomads; and the religious geopolitics of assassins and crusaders.

This book can also be read as a thought experiment focusing on Nietzsche's own thought experiments with time. This experiment consists of reading Nietzsche with regard to questions concerning human time by taking one of his key thoughts, the idea of eternal recurrence, as one of a plurality of temporal modalities. Nietzsche has many things to say about the forms of human temporality, notably political temporality, that would be obscured by prematurely folding them into the thought of eternal recurrence. Recall just a few of the many forms of time he evokes in the titles of books, chapters, and aphorisms: birth and death, the unmodern (*unzeitmässig*), dawn and twilight, great events, noon and great noon, the visionary moment or twinkling of the eye (*Augenblick*), midnight, philosophy of the future. This list could be extended and supplemented with those temporal experiences or constructions that are the subject of Nietzsche's polemics and critiques: the last human, world-history and its perverse cousin the *Weltprozess*, slow suicide (of life in the state), nihilism (as a rejection of futurity), the debt/guilt complex (*Schuld*) which involves mortgaging the future, the cultural philistine's fetishization of the present (again an incomplete catalogue). This experiment in reading Nietzsche on earth and its times draws on suggestions from phenomenological explorations of time, which have always tended toward recognizing a plurality of temporal modes. It finds inspiration in occasional pointers by scholars like Ernst Kantorowicz, who observes in his monumental study *The King's Two Bodies* that "the Middle Ages were perhaps more aware than we are of the various categories or measures of time."[32] Kantorowicz's treatise is "a study in medieval political theology" that attempts to articulate another plural complex of temporalities: the finite time of mortal bodies, the perpetuity of kingship, the time of life in the earth's fallen realm, the king or state's role in postponing the coming of the Antichrist and the end of time.

Nietzsche, I argue, was struggling to free himself and the philosophical tradition from the remnants of political theology (so as I will argue, we should not be surprised that he came to think of his *Antichrist* as either the first book or the whole of the *Transvaluation of Values*). In thinking about Nietzsche's attempt to twist free of political theology, it is essential to realize that he did not abandon his study of theology after his first university year (when he became a student of philology). One indication of this is his long association with Franz Overbeck, professor of theology at Basel, Nietzsche's housemate there, and his most constant correspondent from the Basel time on. Scholars, especially those outside the German language orbit, have been largely unaware of the philosophical (as contrasted with the biographical) significance of the Nietzsche-Overbeck relationship. This is understandable because only since the 1990s has the vast bulk of Overbeck's writings on Christian theology (and what we now call political theology) from earliest times to his own become generally available. In chapter 6, I explore how the shared concerns of the two friends illuminate Nietzsche's critique of world-history and his contrast of world and earth.

Lest readers think I go too far in de-emphasizing the thought of eternal recurrence, I ask them to recall the following: in the August 1881 notes where Nietzsche records his personally epochal experience of the idea in Sils-Maria "six thousand feet beyond man and time," he asks what those who have incorporated (*einverleibt*) the thought are to do with the rest of their lives. He suggests a long political incorporation of the thought, a task possibly for millennia. "We" who have personally incorporated the thought are to teach it, instilling it in the earth's people. Nietzsche sketches the political consequences yielded by the differential responses to the thought's taking hold on a large scale. These range from joyful empowerment to life-threatening dejection, involving massive changes in human life. Rather than folding Nietzsche's political thought into the supposed master thought of recurrence, we will try the experiment of thinking about earth, great events, and great politics by explicating these concepts in their interrelations, their context in Nietzsche's time, and some of their resonances for ours.

CHAPTER 2

Unmodern Thinking: Globalization, the End of History, Great Events

The Christian world is the world of completion; the grand principle of being is realized, consequently the end of days is fully come.

HEGEL, *Philosophy of World History*

Not to cut the Gordian knot, as Alexander did, so that its ends fluttered to all the corners of the earth, but to tie it again—this is now the task. I recognize in Wagner such a counter-Alexander: he unites what was separate, feeble, and inactive.

NIETZSCHE, *Richard Wagner in Bayreuth*, 4

The West has come to encompass the world, and in this movement it disappears as what was supposed to orient the course of this world.

JEAN-LUC NANCY, *The Creation of the World, or Globalization*

In the early 1870s, in the aftermath of the Franco-Prussian War, Nietzsche confronted a German cultural world comparable in some ways to that of the United States and its allies in the 1990s, following the collapse of the Soviet Union and the apparent success of the first Gulf War against Iraq. In both cases, a victorious nation congratulated itself on a newly enhanced position of power and took its victory as an obvious sign of its cultural superiority. Bismarck presided over the founding of the German Reich at Versailles and Bush 41 proclaimed the emergence of a "new world order." In Germany, professors and pundits celebrated the event, taking it as a sign of national cultural superiority. In the United States (and to a somewhat lesser extent among its allies), the end of the Cold War evoked a new confidence in the superiority of Western liberal economics and representative democracy. A new era of world commerce, transportation, and communication was hailed as "globalization" that would have universal benefits. Prevailing opinion saw the radical politics of the 1960s and the economic crisis that marked the transition from the Carter to the Reagan presidency as unfortunate episodes that had now been definitively surmounted. About twenty years later, the United States and much of the West were struggling to come to terms with a new and deeper economic meltdown, terrorism and the rise of radical Islamism, the disastrous invasion and occupation of Iraq, a problematic attempt at nation-building in Afghanistan, and the blatant violation of once-cherished fundamental legal guarantees of privacy, rights to a fair trial, habeas corpus, and prohibition of torture. For Germans, the reversal of fortune proceeded more slowly. It was about forty years after the foundation of the *Reich* that it was destroyed in World War I, and by the 1940s it was possible to speak of this epoch of German history under the rubric "from Bismarck to Hitler."[1] Nevertheless, what many consider the first global economic crisis hit Europe and Germany in 1873 and interrupted postwar self-congratulation.

At the beginning of the Bismarck era *Reich*, the young Nietzsche had a highly complex relation to German nationalism and its ideologists. As a Saxon, he experienced the rise of Prussia with some ambivalence, admiring its military and diplomatic success while skeptical of its ultimate cultural and political implications. Nietzsche served in the German army as a volunteer (suffering various illnesses and injuries) but retained the stateless condition resulting from his recent move to Switzerland as a professor at Basel. He saw it as a time when distinctions between journalism and philosophy were rapidly eroding, with theorists ambitious to become public intellectuals competing to provide satisfying accounts of the time. Nietzsche saw this as the age of the newspaper and the triumphalist German state. He was intensely critical of Hegel's philosophy of history, which he saw in both its original form and its transformations as the ideological core of what was problematic about modernity. Timely or "modern" thinking was at a fever pitch. The young classics professor had recently published his first book, *The Birth of Tragedy*, which held up the tragic age of the Greeks as a mirror in which to observe contemporary Europe's shallow optimism, rationalism, and faith in progress. That book was scorned or ignored by his scholarly peers, who saw it as the work of an enthusiast, entranced by Wagner's "music of the future," which Nietzsche placed on a level with Aeschylus and praised as a great cultural reawakening. On the outs with his scholarly community, Nietzsche turned to writing the series of *Unmodern Observations*, in which he aspired (to speak in recent jargon) to write "crossover books" that could establish him as a "public intellectual." The four published essays, drawn from a list of thirteen that Nietzsche proposed in his notebooks, attack postwar Germany's cultural philistinism, its subordination of life to history, and its capitulation to the journalistic mind (the unmodern [*unzeitmässig*] is opposed to the all-too-timely time [*Zeit*] of the newspaper [*Zeitung*].) The fourth *Unmodern* is a paean to Wagner, appearing in the year of the first Bayreuth festival, and celebrating the advent of what Nietzsche then wanted to believe was an epochal cultural awakening (as his notebooks from the time show, he could not altogether repress his hesitations).

Where possible, I attempt to minimize reliance on biographical and psychological approaches to Nietzsche's thought, while seeking to be informed by historical and political context. Yet it is worth noting that the first steps in Nietzsche's taking up the project of the *Unmoderns* emerged from his interchange with Wagner. The maestro was disappointed that Nietzsche failed to visit at Christmas 1872. There was an unsatisfactory Easter visit the following spring. It was then that Wagner spoke of D. F. Strauss's book *The Old Faith and the New*. Nietzsche began to form the plan of his polemic against Strauss, which was conceived as a present for Wagner's birthday. Since the series breaks off with the fourth installment on Wagner, there is a knot to be untied here, although I will argue that the solution is seen most productively as a conflict in Nietzsche's thinking about history, modernity, and the event, rather than a merely personal drama.

Nietzsche's unfinished series of four *Unmodern Observations* (1873–76) are not only critiques and evaluations of prominent thinkers and artists; they are also explorations of what it means to think or create in relation to one's time, whether as advocate or antagonist. Written in the aftermath of the Franco-Prussian War and the foundation of the German *Reich* by Bismarck, these essays are polemics against cultural philistinism and historicism, personified in the first two essays by Strauss and Hartmann. The third and fourth celebrate Schopenhauer and Wagner, depicted as heroic figures offering bracing alternatives to the smug self-satisfaction of Nietzsche's German contemporaries, who thought they had arrived at something like the end of history. *Unzeitmässige Betrachtungen*, the series title, has frequently been translated as *Untimely Meditations*, and also as *Thoughts Out of Season*. The unmodern (*unzeitgemäss*) is also that which is opposed to the journalistic (*zeitungsgemäss*), so among the many translations proposed for Nietzsche's series, we could consider the somewhat clumsy *Nonjournalistic Reports*. Most recently it has become *Unfashionable Observations* (in the now standard Stanford translation). I prefer *Unmodern Observations* (William Arrowsmith's version) because these essays should be read as opposed not only to the fashion of their own day, but to the implicit and sometimes explicit principle of modernity, according to which history is a progressive, sequential development in which each era subsumes its predecessors. In the specifically Hegelian conception of modernity, one of the declared targets of the series, the modern era incorporates, surpasses, and comprehends all of its predecessors—they are, in the jargon, *aufgehoben*.

These reflections, then, are anti- or postmodern, although the latter term could tempt us to think of the postmodern as a distinctive period that follows the modern. That would simply be to repeat the basic modernist practice in its construction of history. The problem with any locution that would employ the "post" prefix is that it invites a modernist reading; we might experiment (as I have done elsewhere) with a deliberately oxymoronic and ugly term such as "post-periodization" to suggest what is at stake.[2] It might also help to consider a free English paraphrase like *Thinking Off the Clock*. Nietzsche sees his contemporaries being seduced by the idea that humanity (or more specifically its European or German form) has reached a pinnacle of accomplishment, realizing what was only potential in its past. The same attitude is ingredient in the section of *Twilight of the Idols*, "Skirmishes of an Unmodern Thinker" (*Streifzüge einer Unzeitgemässige*), whose attack is directed not only at celebrated figures or "idols" of modernity (e.g., George Sand, Thomas Carlyle, Ernest Renan, and others) but at the modern as a way of construing experience and history. For Nietzsche, living virtually in pre-Socratic Greece and actually in modern Europe, time was out of joint.

The objects of Nietzsche's critique can be usefully compared to more recent "end-of- history" theorists, who, like David Friedrich Strauss and Eduard von Hartmann, targets of Nietzsche's first two essays, share a Hegelian inspiration. In the 1930s, Alexandre Kojève gave a series of lectures in Paris, attended by thinkers like Sartre, Merleau-Ponty, and Lacan, which were published after the war as *Introduction to the Reading of Hegel*. Kojève understood the meaning of history in terms of the Hegelian dialectic of recognition, as outlined in the *Phenomenology of Spirit*, especially the chapter "Lordship and Bondage (*Herrschaft und Knechtschaft*)." If the desire for recognition is fundamental to humans, taking precedence finally over the desire for mere objects (like food), then conflict is inevitable as they risk their lives to demonstrate that they prize courage and honor more than mere life. The universal struggle for recognition, if it should ever transcend a Hobbesian war of all against all, will lead at first to an order in which there are lords and bondsmen (or masters and slaves), in which some accept a subordinate status in return for their lives being spared, while a ruling class obtains the honor of being recognized for their specifically human virtues. This leads to a fundamental instability, for the lords or masters are recognized by bondsmen or slaves, whose recognition must be valueless, since they have abandoned their own quest for honor. At the same time, the subordinate class, through the work imposed on it by its masters, has the opportunity to learn how to shape itself through that work and so to take up the struggle again in new forms.

Kojève's metanarrative of human history proceeds by exploiting this instability in an inspired, Marxist-accented, reading of Hegel, tracing the transformation of the struggle for recognition through such social forms as ancient slavery, feudal hierarchy, and modern capitalism. By the 1950s, Kojève was convinced that the struggle had in principle reached its conclusion. He saw capitalism and socialism as agreeing, despite all their differences, that the only viable society was one that afforded universal recognition to all of its members. The two competing ideologies were simply offering different versions of the content of this universality, both self-described as forms of democracy, one in the form of liberal capitalism, while the socialist path was concerned to establish concrete conditions for human equality. With the rapid disappearance (so Kojève thought) of earlier forms of political claims to legitimacy, whether theological, nationalist, or racist, all that remained was to work out the details of a universal order. As if to underscore the point, Kojève turned from philosophy to administration, as an executive of the emerging European Economic Community, giving his attention to adjusting exchange rates and the price of commodities rather than the exegesis of Hegel's texts. He speculated that history had, in principle, come to an end with the general acceptance of universal human recognition as the basis for society. He wondered whether in such a world human beings were reaching a final state in which the conflict that had been the fundamental motor of history was no longer operative. This might be the era, then, of Nietzsche's "last human," with humanity no longer driven by the struggle for recognition, and satisfied as members of well-organized communities. In a few intriguing but undeveloped notes that Kojève added to his magnum opus after a trip to Japan, he saw one possibility for the continuation of the struggle in another form: humans might turn to aesthetic preoccupations like the Japanese tea ceremony, in which "snobbery," the desire for purely symbolic differentiations in matters of style and art, might take the place of the battles and class conflicts that had hitherto ruled history.

Following the collapse of the USSR around 1991, some analysts came to argue that Kojève was essentially correct. In *The End of History and the Last Man* (1992), Francis Fukuyama took up Kojève's version of Hegel's theory. In the early 1990s, as much of the West was giddily celebrating the demise of the "evil empire," Fukuyama jumped on the bandwagon to provide a philosophical legitimation for the premature party. Accepting the general outlines of the Hegelian dialectic of recognition, he argued like Kojève that polities that failed to institute universal recognition could now sustain themselves only exceptionally and idiosyncratically, because history had demonstrated the availability and superiority of regimes of mutual recognition. Even more emphatically than Kojève, he identified liberal capitalism as the ideal form of othe democratic society of universal recognition. Fukuyama pointed to the ideological bankruptcy of late converts to the liberal democratic way, like Chile and South Africa, noting that their leaders, often recognizing this deficiency, sometimes themselves enabled the reform or transformation of their regimes. While adopting much of Kojève's view and so flirting with the thesis that history was coming to an end, Fukuyama went on to consider the question he had left outstanding as to whether humanity at the end of history would be reduced to the status of Nietzsche's "last humans." He staged a virtual dialogue between Kojève-Hegel and Nietzsche by adding Plato as a third (or fourth) persona, suggesting that the force driving the struggle for recognition is Plato's *thymos*, the aspect of the soul concerned with honor, anger, and pride. In the *Republic*, Socrates maintains a tripartite theory of the soul, said to consist of reason, spirit (*thymos*), and desire. Not only can human beings be best understood in terms of which of these is dominant for them, but polities also have a deep structure depending on whether their ruling element is reason (philosophy), spirit (military honor), or desire (money making). Fukuyama's idea is that the pursuit of honor, glory, and distinction will be generally decoupled from principled political conflicts, while continuing to express itself in sports, art, and scholarship.[3]

Nietzsche's first two polemical essays, on David Strauss and on historical thought, are both, in somewhat different ways, stinging critiques of end-of-history thinking; both pamphlets identify Hegelian philosophy as a crucial component of this thought, thus anticipating Kojève, Fukuyama, and others. The two versions of the theory that Nietzsche considers, Strauss's and Hartmann's, arise from interpreting a teleological conception of history in terms of two variant theories of human desire, one substantially following Hegel's notion of historically cumulative recognition (Strauss), while the other (Hartmann) sees history as a sequenced series of self-understandings that progressively reveal the necessary frustration and failure of the desire for happiness or self-satisfaction (a historicizing of Schopenhauer). Strauss's version is comic, because it celebrates an attained unity arising out of conflict; Hartmann's is tragic, depicting humans as repeatedly attempting the impossible until overtaken by necessity. In each case, Nietzsche's response is satiric and parodic.

His critique of these ambitious metanarratives proceeds on a number of levels, including the demonstration that the stories they tell are just that, stories, which can be deflated by reframing them through an ironic grid. That such deflation was in order is evidenced by the fact that these two books were philosophical best sellers that offered contemporary perspectives on the meaning of history, with each going through multiple editions. Each offered an early version of what has come to be an established genre: theories that offer a sweeping view of human history and a diagnosis for the present and future that relies heavily on a "scientific" foundation (frequently Darwinian, as in these two). Since Nietzsche's own thought has some resemblance to work in this genre, when considered from a sufficiently general perspective, it is important to see how much he rejects in Strauss and Hartmann. Specifically, he sees these "end-of-history" worldviews as cultural diseases that require a hygienic antidote. Because they ignore—or worse, condemn—the genius, the exception, and the lucky hit, seeing them as nothing more than minor and dispensable variations in relatively uniform populations or masses, Nietzsche charges that they systemically blind themselves to sources of value and change, becoming ideologies of what he will call the last human.

THE END OF HISTORY: NATIONALIST TRIUMPHALISM

Nietzsche later referred to "David Strauss, the Confessor and Writer" (1873) as the duel or assassination attempt with which he entered society (*EH*, "Unmodern Ones," 1–2). Readers have tended to neglect the Strauss essay, while giving their attention to the second *Unmodern* on history. Yet in explaining "Why I Write Such Good Books" in *Ecce Homo*, Nietzsche devotes much space to the first in the series. Celebrating his warlike entrance into the world of intellectual polemics, he quotes a saying of Muhammad's, "paradise is under the shadow of swords," suggesting his joyful participation in a *jihad* of the spirit.[4] It is fitting that Nietzsche would choose a non-European insignia in his later reflection on his war against modernity. As late as *The Antichrist* (1888), Nietzsche recalls his youthful admiration for Strauss, author of the pathbreaking *Life of Jesus* (1835): "like every youthful scholar," he writes, "I took great pleasure in the work of the inimitable Strauss, read at the shrewd and plodding pace of a refined philologist" (*AC* 28). By philology Nietzsche understands, briefly, a rigorous, fine-tuned, and vigilant attention to differences in language, meaning, and tone, informed by knowledge of context. Exemplary is the work of the pioneer philologists working on the "Homeric problem," who saw a supposedly unified text as a palimpsest of multiple texts, authors, and traditions. Strauss applied analogous critical principles to the New Testament. Nietzsche's best-known philological exercise is probably his historically and linguistically grounded explanation of the distinction between two value systems: "good and bad" and "good and evil" in *On the Genealogy of Morality* (anticipating Jacques Derrida's lapidary characterization of deconstruction as "always more than one language"). By 1872, when he published *The Old Faith and the New*, Strauss had become an advocate of Bismarck's *Reich*, claiming that Germany's victory in the Franco-Prussian War, and the consolidation of the *Reich* under Prussian hegemony, represented a cultural and not merely a military victory. While he had always moved within a Hegelian orbit, Strauss's "Bismarck book" comes very close to celebrating the German present as the ultimate realization of national and human potential. Nietzsche denounced this attitude as "cultural Philistinism," a term which he claims to have introduced into German.

Strauss's treatise begins with what sounds to most contemporary readers like an antiquated and parochial set of religious disputes, first between Protestantism and Catholicism, and then within these branches of Christianity themselves. It was the time of the *Kulturkampf* or "culture war," mainly instigated by Bismarck, who sought to minimize the power of the Catholic Church. As Strauss notes, the foundation of the *Reich*, declared by Bismarck at Versailles, was roughly simultaneous with the Vatican's declaration of papal infallibility, threatening to strengthen the church's political-ecclesiastical power in Europe. This sets the stage for four questions from Strauss: "Are We Still Christians?," "Have We Still a Religion?," "What Is Our Conception of the Universe?," and "What Is our Rule of Life?" As Nietzsche quickly points out, central to these questions and their answers is Strauss's conception of the "we," the community assumed in these questions. He maintains that Strauss has loaded the deck by implicitly modeling this "we" on the self-satisfied German citizen, whose culture is drawn from the newspapers, whose religion is an ethical ghost of Christianity, whose faith in progress can envision nothing beyond further sophistications of communication and technology—not far from Zarathustra's last human. Strauss could be the last human's philosopher, declaring, "Yes, history has a meaning and it is us." Strauss claims to be ultimately timely, to have understood the fulfilled meaning of time, indeed, as part of his "we," to help constitute that meaning. Nietzsche's response is untimely, unseasonable thinking, or thought that is "off the clock."

Nietzsche finds humor in some of Strauss's "fanatical opponents" identifying him with "the beast from the abyss" (as in the Christian *Apocalypse*), but Strauss invites such identifications insofar as he approaches the idea of a non-Christian, this-worldly end of history (*UO* I.8).[5] Since Nietzsche's own political thought concludes with a book called *The Antichrist*, it is worth noting what seems to be his first reference to this diabolical figure associated with the Christian end-times. Nietzsche's study of Strauss appeared in the same year as another polemical pamphlet by the Church historian Franz Overbeck, his lifelong friend and, at that time, his housemate in Basel. In the last sections of Overbeck's essay, Strauss's *The Old Faith and the New* becomes a target of pitiless satire and irony. The pamphlet *How Christian Is Our Present-day Theology?* argues that discursive, rational theology is completely antithetical to the original spirit of Christianity, which involved the apocalyptic expectation of an imminent end of the world, and disdained any attempts to justify faith through rational argument.[6] Eventually Nietzsche will ask what the true name of the Antichrist is, indicating that only with his own (Nietzsche's) thought is a philosophy of the Antichrist possible, and finally sign himself as Antichrist. Later we will see that these are not simply hyperbolic, extravagant expressions (sometimes attributed to supposed madness) but are grounded in Nietzsche's knowledge of Christian texts and traditions, including a tradition of political theology that has recently re-entered political philosophy through discussion of Carl Schmitt, Giorgio Agamben, and others (see chapter 6).

While Strauss mentions Hegel only fleetingly in the book Nietzsche pillories, presenting his own ethics and politics as a form of humanist universalism rooted in the Stoics and Kant, Nietzsche rightly discerns that Strauss's "we" is a Hegelian first-person plural. He cites Strauss's claim that God (who is pretty much identical with human reason at this point) "shows us that while chance would be an irrational master of the world, necessity, i.e., the chain of causes in the world, is reason itself." Nietzsche comments that Strauss has performed a sleight-of-hand here, conflating scientific law with ethical norms, attempting to absorb Darwinian evolution within a Hegelian story of moral progress, observing that "only the 'we' can fail to recognize this piece of surreptitiousness as such because they were raised in this Hegelian worship of the real as the rational, that is to say in the deification of success" (*UO* II.6). In his notebook sketches from this time, Nietzsche wrote, "The Hegelian 'world process' culminated in a fat Prussian state with a capable police force."[7] And indeed, the entire fourth part of Strauss's treatise, which purportedly provides an answer to "What Is Our Rule of Life?," has a vulgarly Hegelian logic. Strauss begins with a universal humanism, which supposedly survives the dissolution of the old faith (Nietzsche will wonder why it should survive the death of God). The universality here seems to be a somewhat weaker, less robust version of Hegelian recognition. This humanism itself is determined by a law of *kind* or species, and so actual norms and forms of practice will be based on the most immediate kinds in which humans find themselves associated: tribal, racial, ethnic, linguistic, and national. Naturally this undermines its claimed universalism. Of these it is the national that Strauss sees as embracing or superseding other forms of kind, but as barring the formation of any genuine supranational political order. Such an order would have to prohibit war, and Strauss (like Hegel and many European contemporaries) takes war to be not only inevitable but a necessary accompaniment of progress and human self-realization. Accordingly, his "we" becomes the voice of Germans, who see themselves as having demonstrated their cultural superiority by military victory and territorial expansion (as in the sleight-of-hand discussed earlier). A strong national identity is necessary for a state and is the only source of patriotism, as Strauss (in the aftermath of the US Civil War) points to the example of republics, whose people lack homogeneity and so, incapable of true patriotism, dissolve into social atoms.[8]

Strauss does not hesitate to argue specifically for the priority of monarchy, with its "necessary mystery," over a republic, and against universal suffrage, which threatens democratic excess, social anarchy, and the outrages of the Paris Commune. He offers as a telling sign that republics like the United States and Switzerland produce no great literature or art (at this time Nietzsche was living in Switzerland, having abandoned German citizenship, and was an enthusiastic reader of the American Ralph Waldo Emerson; he also admired the Swiss Gottfried Keller). Most of all, Strauss is concerned to defend the integrity of the German Reich against the "fourth estate" of the working class and international organizations, especially the International and the Jesuits: "Only in its natural division into nations may mankind approach the goal of its destiny; he who despises this division, who has no reverence for what is national, we may fairly point to as *hic niger est*, whether he wears the black cowl or the red cap."[9] In this context, the old classicist's quotation from Horace can be read as having racist as well as anti-Catholic and antisocialist overtones.

Strauss's criticism of republican government and democracy, as well as his insistence on the necessity of monarchy to provide a principle of national unity are almost identical with Hegel's views. The most significant point of difference is his dropping Hegel's requirement that the state embody a national religion. Remember that Hegel defines world-history as the history of states and eliminates all prestate or nonstate migrations or wanderings of peoples as outside the sphere of real history.[10] When Hegel famously describes world-history as a slaughter-bench, he is not speaking of the violence of some state of nature (which is prehistorical) but mostly about the destruction of republics (Greece, Rome, the Italian city-states, the first French republic). Nietzsche is specifically suspicious of the apparently simple term "world (*Welt*)," which figures so strongly in Hegel. The latter insists that world-history is the history of states and begins only with the formation of states. In his notebooks Nietzsche contrasts this conception of world with that of earth, anticipating Zarathustra, who speaks of the direction of the earth, not of the world.[11] Hegel's explicit examples of world-historical figures—like Caesar, Alexander, and Napoleon—are those whose mission was to transform republics into empires.

Even Strauss's description of the United States as a spurious union echoes a specific diagnosis that Hegel offers in his *Lectures on the Philosophy of World-History*.[12] Hegel implies that the United States is not really a state and has only a starkly contractarian and atomistic parody of a genuine constitution. Forty years or so later, Strauss observed the US Civil War and its aftermath, and the Hegelian argument appeared to have been confirmed. Hegel had to explain how this simulacrum of a state exists—after all, the real is the rational, and the rational is the real—and so argued that its territorial expansion served as a safety valve through which the excesses of a "state" not grounded in a *Volk* or even given substance by monarchy and religion can continue. He doesn't discuss the question of just what might develop when the limits of expansion are reached or when, as it happened, the North and the South came into conflict about the extension of slavery. That crisis erupted around the very safety valve of internal emigration and pioneering that, according to Hegel, made this strange substitute for a state temporarily possible. Hegel might have seen the US's western move to Hawaii and Alaska as an understandable extension of the westward movement of world-history. He could have gone on to explain the Alaskan secessionist movement as an indication of the impossibility of the contractarian state, and he would have understood its affiliation with an apocalyptic and territorial form of Christianity that reverts to what he would consider prehistorical forms of animism and belief in witches. With the division into red states and blue states, and looming conflicts over energy, water, immigration, and the fundamentalist social agenda, Hegel would wonder whether the experiment of the self-designing constitutional republic without a religion could be expected to continue indefinitely. While this could be a plausible Hegelian story, we can also deploy the experiment of a continuing secular, multicultural republic as an incentive to interrogating Hegel's conception of world-history. Nietzsche, I'll argue later, questions most of this Hegelian rejection of the possibility of a polity without monarchy, religion, or ethnic nationalism, most notably in his analysis of Europe's desire to "become one" (in chapter 8 of *Beyond Good and Evil*, which aims at thinking beyond "Peoples and Fatherlands").

Looking back after the stretch of German and European history "from Bismarck to Hitler," it is not difficult to see the prescience of Nietzsche's response: "a great victory is a great danger," more specifically the danger of understanding a military or political victory as a cultural one, thus risking "*the defeat, if not the extirpation, of the German spirit for the benefit of the 'German Reich'* " (*UO* I.1). Nietzsche remained suspicious of the *Reich* and developed ever more critical accounts of the kind of nationalism typified by Strauss. We may take one more step and note the parallel between Strauss's triumphalism in the time of the *Reichsgründung* and the end-of-history ideology that accompanied US emergence as the world's sole superpower in the early 1990s; the latter coincided with the announcement (by George H. W. Bush) of a "new world order." While Strauss's geopolitical vision is decidedly Eurocentric, and relatively unaffected by the awareness of growing globalization that Marx and Engels had already noted in *The Communist Manifesto,* he hailed Germany's rise in Europe in much the same spirit as Fukuyama greeted the apparent hegemony of the United States and the capitalist democracies following the collapse of the USSR. Fukuyama's vision of the future of liberal democracies has so far not been validated by events. He suggested that the first Gulf War had definitively shown that nations would no longer pursue war for oil and natural resources, a claim cast into doubt by the 2003 Iraq war and other conflicts unfolding in that area of the globe.[13] Fukuyama may or may not be right in his claim that a "liberal democracy that could fight a short and decisive war every generation or so to defend its liberty and independence would be far healthier and more satisfied than one that experienced nothing but continuous peace."[14] Fukuyama wrote this just a year or so after the United States ousted Saddam Hussein from Kuwait, but well before a generation had passed, the United States declared a global war on terror, whose acknowledged fronts are the ongoing US wars in Afghanistan (2001–2016?) and Iraq (2003–2011, but now morphed into a war on ISIL), neither of them "short and decisive," with their bloody internecine and global consequences still unfolding.

The self-satisfaction of Strauss's "we," Nietzsche argues—and we could add Fukuyama's liberal capitalists here—leads both to the practice and the theoretical defense of stagnation:

> All seeking is at an end is the motto of the philistines . . . they devised the concept of the epigone-age with the object of obtaining peace and quiet and so as to meet every uncomfortable innovation with the condemnatory verdict "epigone-work" . . . Through historical awareness they saved themselves from enthusiasm . . . the sole proviso [with regard to cultural creativity] was that

> everything must remain as it was before, that nothing should at any price undermine the "rational" and the "real," that is to say the philistine (*UO* I.2).

Kojève, perhaps even more insistently than Hegel, spoke of the goal of human activity as satisfaction, describing its final goal as satisfaction of the desire for recognition. Nietzsche sees Strauss's "we" as satisfied in only a weak sense, because its demands are so low; therefore, he argues, the "we" itself must be weak. Anticipating later claims (made forcefully in the *Genealogy*), he brings to light the apparent paradox that the weak can hold power by outnumbering the strong; they betray their weakness by an extreme defensiveness and inability to brush off criticisms of their character (*UO* I.3).

THE END OF HISTORY: APOCALYPTIC GLOBALIZATION

Nietzsche's next pamphlet, *On the Advantages and Disadvantages of History for Life*, has become the most read of the *Unmoderns*, although Nietzsche had great reservations about this text. He now turns his guns with gleeful abandon against a passive, tragic end-of-history scenario. Here the diagnosis of modernity is carried out in medical and geographical terms, that is, in a rhetoric designed to frame its subject clinically and spatially rather than temporally. Considered as a malaise, modernity is inflamed and empowered by a conception of history and historical education that assumes the inevitability of modernity itself. To be modern is to be timely (*zeitmässig*), and so to know one's position in time. It is necessarily to be a latecomer. The most extreme form of the modern historical malaise is the conception of history as a science, which Nietzsche attributes to Hegel and the Hegelians. In this view, the core of modern culture is a historical education that supposedly allows the internalization of everything of human significance. It offers "the illusory promise that it is possible to sum up in oneself the highest and most noteworthy experiences of former ages, and precisely the greatest of former ages in a few years" (*UO* II.10). The science then takes itself to be the outcome of the historical process, therefore a preeminent form of consciousness that lies at the end of a long road: "History become pure, sovereign science would be for the human race a sort of conclusion of life and a settling of accounts with it" (*UO* II.2). Nietzsche sometimes describes this overtaking of life by scientific history as a form of self-mummification (*T*, "'Reason'" 1).

Speaking as cultural physician and climatologist, Nietzsche offers three possible alternatives to the stultifying regimen of the historical science that seemed the inevitable ideology of the Bismarck era. Each of these will work only in a certain context: "All living things require an atmosphere around them, a mysterious misty vapor . . . every nation, too, indeed every human being that wants to become mature requires a similar enveloping illusion, a similar protective and veiling cloud" (*UO* II.7; cf. *UO* II.2). Monumental history, seeking inspiration and example in great deeds of the past, is suited to those who act and strive, but is liable to abuse by those who envy and belittle present possibilities of action by invidious comparisons with the past. History in the antiquarian mode provides ground and continuity for those who preserve and conserve, yet can degenerate into a fetishism of tradition (mere love of old "furniture") and so stifle any sense of futurity. Critical history rightly appeals to those who need liberation from suffering; at its best it eliminates barriers to action and frees up energies, while if improperly deployed it becomes simply a blind engine of destruction.

It is not only Hegel's general notion of history as science that provokes Nietzsche's second *Unmodern* essay. The work is also informed by a confrontation with Hegel's specific story about world-history, a story that culminates in the Germanic world, which has discovered that "all are free." Nietzsche offers a diagnosis of "present day Germans" that reads like a reversal of Hegel's account of the Germanic world in his *Lectures* (*UO* II.4) There Hegel argues that the special characteristic of the Germans was their receptivity, a trait he traces back to the wandering peoples who attacked the Roman Empire. Initially it may seem strange that the Germanic world, which will see the full flowering of Spirit and of the state, begins with barbarous, wandering, predatory, *prehistoric* peoples—Goths, Visigoths, and so on. Yet Hegel implies that these groups are no different than any others; no *Volk* enters history until engaged in the process of state formation. Hegel makes German barbarism a virtue, claiming that the Germans' strength was to begin by absorbing and appropriating, unlike earlier historical peoples who begin with an internal development. For Hegel, the very being of the Germanic peoples is their becoming through encounters with the other. They seize Rome and appropriate Christianity almost thoughtlessly. They end up transformed by what they have captured. They are predatory *subjects* who will be reshaped by their *object*. This heritage allows them, through the Reformation and the development of the modern state, to spiritualize the secular; thus they are uniquely suited to confirm Hegel's concept of the true identity as that of identity and nonidentity.[15]

Nietzsche agrees that the Germans are receptive, but describes them as borrowing the fashions of others in dress, culture, and ideas with no sense of form: "The subjectivity of the Germans can be receptive to an exceptional degree: serious, profound, and perhaps even richer than that of other nations; but as a whole it remains weak because all these beautiful threads are not wound together into a powerful knot." Nietzsche sees the Germans as culturally confused, torn between an encyclopedic, subjective, and disorderly interior appropriation of art, histories, and ideas and their crude manner of letting themselves go, dwelling, for example, "in a carelessly inaccurate copy of French convention." The discrepancy seems especially striking in contrast with the self-satisfaction in the wake of the Franco-Prussian War and the *Reichsgründung*. Nevertheless, there is something reminiscent of Hegelianism in Nietzsche's call for true German unity: "it is for German unity in that highest sense that we strive, and strive more ardently than we do for political reunification, *the unity of German spirit and life after the abolition of the antithesis of form and content, of subjectivity and convention.*—" (*UO* II.4). Hegel also values cultures and states in terms of their attaining complex forms of unity, and in the final analysis it is art, religion, and philosophy that are the highest forms of spirit, transcending and including the political state that is their presupposition. Later Nietzsche will considerably relax this standard of unity when he seemingly endorses Europe's desire to "become one" (*BGE* 256). As we will see, this involves accepting something less than a highly structured cultural unity for an individual people, like the Germans, and even the oneness that Europe is said to seek will, on closer inspection, allow for hybridity, nomadism, and cosmopolitanism; the truth of this "becoming one" is in fact plural, what Nietzsche will call "the century of the *multitude* [*Menge*]" (*BGE* 256).

While readers of the *Advantage* are typically alert to Nietzsche's critique of the Germans and to the intense historicism he saw dominating the German intellectual scene, they may be puzzled by the last quarter of the essay, which is devoted to excoriating Eduard von Hartmann, a philosopher little known now but one (like Strauss) well known to Nietzsche's audience (*UO* II. 8–9). If Strauss vulgarizes Hegel's story, and anticipates the end-of-history narratives of Kojève and Fukuyama, Eduard von Hartmann's variation on the metahistorical scheme has a more explicitly globalizing scope and operates with an alternative, Schopenhauerian, pessimistic conception of desire as its agent of historical change. Almost all of Nietzsche's commentators have neglected Hartmann's specific historical and metahistorical claims, which are the object of devastating critique in the concluding sections of his essay.[16] Hartmann, like Strauss, aims at being timely, producing a theory that explains the inevitability of the present. If his notion that history is a meaningful progression of epochs, in which humans try out various strategies for attaining satisfaction, is Hegelian, his analysis of these as a cumulative series of failures, ending in an ultimate failure already taking shape in the present, can be traced to concepts of desire and satisfaction adapted from Schopenhauer. Nietzsche ridiculed the facile Hegelian optimism with which Strauss dismissed Schopenhauer. Now he savagely parodies the translation of Schopenhauer into a historical account purporting to demonstrate the necessity of humanity consciously realizing the fundamental futility of its project of happiness.

Hartmann's ambitious story moves through four great periods, and so was attractive to the nineteenth century's post-Hegelian periodizing obsessions. These epochs follow a traditional analogy between history at large and the standard four stages of the individual life cycle: childhood, youth or adolescence, maturity, and old age. Hartmann implicitly restricts his narrative to the West, apparently accepting the prejudice of his time and place that the East had no real history, a belief that helped to support European imperialism, then on the rise. The story proceeds: In their Greco-Roman childhood, human beings simply seek happiness in this life, and naively enjoy immediate life activities as imaginatively sanctified and perfected by the Olympian gods. Yet such a life eventually disappoints and, Hartmann says, gives way to feelings of boredom, melancholy, and disgust. There is a growing realization of the impossibility of this-worldly happiness. The alternative to such despair is found in medieval adolescence, which places its hopes for happiness in the fiction of immortality, in another world. Such hopes do not bring happiness here and now, and (as Nietzsche will say in his own way), the search for truth ignited by the quest for absolutes leads finally to skeptical disillusion about the reality of the world beyond, on which believers pinned their hopes. Mature manhood (the terms of Hartmann's analogy are unreflectively gendered), that is the post-Reformation Western world or modernity, no longer seeks the immediate enjoyments of childhood or directs its energies toward fantastic adolescent ideals. Courageously shouldering its responsibilities, it surrenders full happiness for itself, finding satisfaction rather in working toward a general progress of civilization that promises happiness to future generations. Now, after several centuries of such effort, Hartmann claims that a general disillusion about the future earthly paradise is setting in. Human beings doubt that they are really becoming happier and note that the civilization of modernity, with its science, industry, and unrestrained economic development, is accompanied by its own discontents, which do not automatically fade away. We are now, it seems, entering the world's disillusioned old age, in which we understand the failure of humanity's earlier attempts at satisfaction. This is the time, for example, of

Schopenhauer's philosophy, which explains the impossibility of human happiness in terms of the basic characteristics of desire. Schopenhauer's Will is restless and insatiable; any temporary satisfaction is followed by melancholy or the return of desire which simply relaunches the cycle of frustration, striving, fleeting satisfaction, and repeated frustration. In its old age, humanity submits itself to the *Weltprozess* that has brought it to this pass and finally becomes aware of its earlier foolishness. Old age prides itself on understanding that process and its insight into the fundamental painfulness of life. It welcomes the process of human extinction.

Hartmann's elephantine treatise *The Philosophy of the Unconscious* purports to show the unconscious dynamics that determine the *Weltprozess*. That there is such a process, a movement in a discernible direction taking precedence over all else, is the crude teleological notion that Nietzsche ridicules ontologically and contests from the standpoint of ethical and political action. If Hegel and Strauss offer accounts of the identity of the real and the rational, a story of how humanity (or at least Germany) attains deep satisfaction, Hartmann opposes that story with the narrative of a *via dolorosa* with its own destined conclusion. Nietzsche applies a scalpel to the virtues of his time, inviting us to question its typical assumption that the historical sense is a virtue. Nonhuman animals are not burdened by a sense of the past or expectations for the future; but every human must come "to understand the phrase 'it was': that password which gives conflict, suffering, and satiety access to humans so as to remind them what their existence fundamentally is—an imperfect tense that can never become a perfect one" (*UO* II.1). Yet if we are condemned to live and think historically, Nietzsche asks, what are the advantages and disadvantages of historical thinking? Here he distinguishes the three forms of the historical consciousness discussed earlier.

Hartmann's philosophy of history, with its strange combination of Hegelian development and Schopenhauerian pessimism, was perhaps as close as the nineteenth century got to producing a theory of the end of history *and* of the last human. In the historical part of his treatise, devoted to showing humanity's gradual ascent to self-knowledge through realizing the impossibility of happiness, whether in religious or secular form, Hartmann explains that the modern age must lead to a world connected by commerce and communication. The white races of the northern hemisphere will skillfully exercise rule over the rest of the world, wherever possible provoking the other peoples of the earth to eliminate each other.[17] However successful these first-world people are at providing themselves with the opportunity for a fully secular happiness, they will finally see that it is not possible to live happily on the earth and that there is no other life to look forward to. Nietzsche took this early theory of globalization, however shallow, to be symptomatic of the last human's possible hegemony over the earth.

An important aspect of the globalization theorized by von Hartmann and others is the fusion of space and time, which minimizes earthly differences and diversity. A recent book on the global standardization of time in the later nineteenth century calls the years from 1875–85 "the decade of time."[18] It was then that the local time that had hitherto prevailed everywhere throughout the earth came to be seen as an obstacle to the requirements of speed, transportation, and communication. In local time, as marked by a sundial, for example, noon is fixed by the daily high point of the sun, the time of "the smallest shadow," as Nietzsche says. But since every twelve miles of the earth's circumference marks a difference of a solar minute in local time, each city would have its clocks set differently. Britain was the first nation to adopt a standardized time in 1850, due largely to the internal expansion and intensity of its rail system. In 1869 the United States completed the transcontinental railway when the Golden Spike was driven in a remote area of Utah. But the cities linked by rail were separated by an anarchic time system, so that a traveler, in making plans, might have to calculate with five or six different local times. In the following years, as Nietzsche was writing and publishing the successive parts of *Zarathustra*, there were a series of international meetings that sought to resolve the dilemmas caused by the collision between the new technologies of speed and communication and the continued plurality of local times. The final result was the establishment of the twenty-four global time zones that we have now, keyed with some variation, to lines of longitude. The standardization of time involved the striation of space associated with the state form, and then with an emerging global system; if the railroads offered Nietzsche some of the nomadic possibilities of smooth space, they also demonstrated to him the ominous power of number, weight, and measure, which as William Blake says, should be brought out only in years of dearth. If the human completes a certain trajectory in a total measuremnt of the earth, then the posthuman (*Übermensch*) could be the agent whose excess (*Übermass*) involves abandoning such limited forms of measure.

From this perspective, Hartmann's construction of the *Weltprozess* (a term Nietzsche finds endlessly amusing) is an abuse even more antithetical to life than any of the defective modes of monumental, antiquarian, and critical history, for it claims that the lesson to be learned from the past, what history is teaching to the contemporary world, is the futility of life itself. When the historical sense dominates "without restraint," Nietzsche says, it "uproots the future," that is, it forecloses any openness to futurity or advent (*UO* II.7). Heidegger and Derrida have attempted to highlight this sense of futurity by emphasizing the sheer unknowability of the *Zu-kunft* or *a-venir*, hyphenating the words to bring out this dimension of meaning in what is to come or arrive. Nietzsche's *Beyond Good and Evil: Prelude to a Philosophy of the Future* (*Zukunft*) should also be read in these terms; it is not prediction or prophecy but an attempt to open thinking to futurity. Nietzsche speaks disparagingly of the *hochmodern* life, which we might translate as "postmodern," urging his contemporaries to think "not of carrying their generation to the grave but of founding a new generation" (*UO* II.8).

While Hartmann professes atheism (like the later Strauss), he is not above invoking God metaphorically, apparently referring to the unconscious *Weltprozess* (internalizing the conflict between "left" and "right" Hegelians, who differed on whether Hegel's theology ought to be read metaphorically). Yet Nietzsche argues with some force that Hartmann's Hegeloid conception of the ages of humanity is simply inherited from Christian medieval narrative, with its apocalyptic scheme of first and last things, culminating in a "fearfully expected judgment" and so infected by a memento mori (*UO* II.8). As such, Hegel's and Hartmann's histories are "disguised theology." In this instance Nietzsche suggests that history must be used to dissolve the problem of history and "knowledge must turn its sting against itself"; that is, by unearthing Hartmann's implicit Christian historical scheme, we can realize how bizarrely tendentious these metanarratives are. Hegel, on this view, is responsible for seriously endangering German education and culture (*Bildung*) by assembling a complex of interrelated themes: the notion that we are all latecomers, that there is indeed a world-process, and that the history of the world that it unfolds is the judgment of the world. Hartmann is then a nihilistic version of end-of-history thinking, one pointing to the possibility of Kojève's and Fukuyama's thought being easily given the same twist in variations of what Zarathustra will call "the last human." Already in the *Unmodern Observations* Nietzsche is aware of the problematic of political theology, which becomes more explicit in the Antichrist theme emerging in his last works. From the standpoint of a certain strain of Christian apocalyptic (revived in the twentieth century by Carl Schmitt), Hartmann's end-time vision of the coming of Antichrist depends on his historical claim that there is no longer a restraining force or *katechon* capable of delaying this destined end. As we shall see, Nietzsche welcomes the weakness or corruption of potential *katechontes*, while transvaluing the unleashed powers that have gone under the name Antichrist, and might be better designated as Dionysus (*BT*, 1886 Preface).

In a delicious Kierkegaardian moment, Nietzsche describes Hartmann's work as a jest, a parody of gigantic proportions, meant to reduce Hegelian philosophy of history to the absurd (*UO* II.9).[19] Surely, he claims with tongue in cheek, this book is a parodic demonstration of the foolishness of contemporary historicism. It thus expresses the ironic consciousness of itself that the age has been developing under Hegel's aegis; the irony becomes cynicism when it construes all of past history as nothing but a "preparation for the everyday use of modern man." Today we should be struck by Hartmann's notion of inevitable globalization and racial extermination—he favors missionary work and commerce as more effective means of eliminating "inferior races" than outright warfare—and by his belief that North America's "republican pyramid" or oligarchy of egoists represents the ultimate form of political organization. Hartmann wants to speed humankind on to its self-consciously Schopenhauerian old age; and he underlines the coincidence of his notion of the direction of the "world-process" with Christian apocalyptics and its notion that we are living in "the last days."[20]

Reading Nietzsche on Hartmann in this time of wars that are often cast as religious wars, Crusades or *jihads*, we note his theological analysis of Hartmann as an end-time thinker inspired by Christian tradition. He detects the "apocalyptic light" that pervades the account of how history comes to an end, an end that is really a philosophical transformation of the apocalyptic "judgment day" (*UO* II.9). Nietzsche's critique of Strauss and Hartmann prefigures the extraordinary convergence of neoliberal capitalist globalization theory and practice with renewed end-time thinking, which came together in the George W. Bush administration, as cynical advocates of US military and economic hegemony played upon Bush's born-again theology to package the "global war on terror," specifically the Iraq invasion, as an irresistible progress of "God's gift of freedom." That some of Bush's advisors (and Fukuyama) were educated by Leo Strauss at the University of Chicago casts an interesting light on this theologico-political complex. Leo Strauss has been interpreted as holding a nuanced and cynical view of the relationship of the religious and the political. In this perspective, the enlightened philosopher is quite aware of the illusory character of traditional religious belief. Nevertheless, the philosopher understands that the population is incapable of this insight and would turn against the wise man if they became aware of the latter's skeptical leanings. Accordingly, the wise man instructs and encourages political leaders in the shrewd use of religion for practical ends. In some cases (like that of Bush 43), the Straussian advisor will see the prudent course to be steering leaders by making use of the illusions they share with a critical component of the voters so as to get on with the business of economic and military expansion.

Hartmann has a robust theory of globalization. First, he sees the direction of history in the expansion of global commerce, transportation, and communication, all with a decidedly imperialistic cast. Second, his very emphasis on "world" and "world-process" incorporates a notion of global totality and inevitability that anticipates more recent accounts. In calling for the "total surrender of the personality to the world-process," Hartmann obscures what Nietzsche thinks are better terms of analysis, namely the human and the earth. "If only one did not eternally have to hear the hyperbole of all hyperboles, the word world, world, world, since after all, if we remain honest, everyone ought to speak of human, human, human!" To call for total surrender to the world-process is to give humans the personality of the earth-flea (*Erdfloh*), a metaphor that appears again when Zarathustra describes the last human: "For the earth has now become small, and upon it hops the last human, who makes everything small. Its race is as inexterminable as the earth-flea; the last human lives the longest" (*Z* P5).

Nietzsche took this early theory of globalization, however shallow, to be symptomatic of the last human's possible hegemony over the earth, a specter that haunts him throughout his work and that Zarathustra cites as the greatest obstacle to fully incorporating his own thought of eternal recurrence (*Z* III.13. 2). He also began to develop his own account of geographical and historical time and space. Nietzsche is not only "untimely" in his critique of the idols of his time but is concerned above all to challenge modernity's sense of historical time. Modernism in this sense is the attempt to construct a metanarrative issuing in a utopia projecting the hegemony of technology, technocracy, and Western imperialism. The program of Nietzsche's postperiodization (the only significance I can find for the concept of postmodernism[21]) consists in elaborating alternative notions of time and futurity.

Nietzsche argues that geography takes precedence over history in contextualizing human action; from this perspective, the Eurocentric limits of globalization theory become apparent. The ultimate object of political and historical analysis, site of life, and field of contest is best understood as *earth* rather than *world*. History, unlike geography, leaves an opening for transcendence in a supposed final meaning. In response to Hartmann's imperialist, capitalist, racist, apocalyptic globalization scenario (pandering to Christianity), with its figure of the last human and its promised end of history, Nietzsche questions both his underlying historicism and his metaphysics, that could construct a "world-process" leading us inexorably to these ends. His alternative account of human action explains the contrast between the tendencies produced by the exaggerated historicism he has been criticizing and the life of people inhabiting an earthly horizon. Nietzsche claims that a certain kind of historical thinking is being put to ideological uses to promote an egoism that we might now call narcissism:

> *the aberrations of the historical sense from which the present time suffers are intentionally furthered, encouraged and—used.*
>
> And they are used against youth in order to train them to that manhood of egoism which is everywhere aspired to . . . A certain excess of history is capable of all of this, we have seen it: through a continuous displacement of horizon-perspectives, through the elimination of an enveloping atmosphere, it no longer permits humans to feel and act *unhistorically* (*UO* II.9).

Nietzsche's terms "horizon" and "enveloping atmosphere" suggest an alternative way of construing human action, one in which the totalizing concept of "world" has been put out of commission and replaced by the conjunction of human and earth. "World," as used by Hartmann introduces a conception of total meaning or totality. *Weltprozess* can be read as a modification of Hegel's *Weltgeist*. "Process" substitutes for "spirit" because Hartmann's is a philosophy of the *un*conscious, an attempt to meld a Darwinian account of evolution with a Schopenhauerian theory of impossible desire. This combination is then forced into a Hegelian frame, retaining the sense of unity that is associated with "world." Recall too that Hegel's world is a rather exclusive club of states, excluding wandering and migratory peoples. Hegel regards those modern polities (like the United States) that lack monarchy and established religion as shaky candidates for provisional admission, and he sees both Islamic and Jewish states as long-surpassed historical curiosities because of their uncertain relations to national territories.

TIMELY, ALL-TOO-TIMELY: PHILOSOPHY'S DESCENT TO JOURNALISM

Readers usually notice that *Schopenhauer as Educator* is only glancingly concerned with what are generally taken to be Schopenhauer's central ideas, such as the opposition and play of will and representation; rather, it appears to be devoted to his exemplary status as an individual thinker who heroically resists the trends and temptations of his time. Stanley Cavell has argued that Nietzsche's Schopenhauer essay cannot be read—as it was, for example, by John Rawls—as proposing an absolute gap between a small elite capable of a perfectionist ethics and all the rest of the human race. Cavell notes that Nietzsche says "culture is the child of each individual's self-knowledge and dissatisfaction with himself," suggesting that no one is excluded from the path of perfectionism (*UO* III.6). That Nietzsche is paraphrasing Emerson here makes Rawls's reading even more doubtful.[22] If Hartmann abused Schopenhauer's theory of desire, foreclosing any futurity beyond submission to the *Weltprozess*, Nietzsche now invokes the great pessimist to show that a dedicated and rigorous pursuit of *Bildung* (culture) is still possible in a mediatized age, through focusing on his predecessor's treatment of the iconic figures of the philosopher, the artist, and the saint. A major part of this demonstration is Nietzsche's emphasis on Schopenhauer's distance from the culture and institution of the state; he puts him to totally different uses than did Hartmann, who attempts to trace the world-historical consequences of a growing acceptance of a pessimistic worldview. In developing Schopenhauer's critique of the state and state-oriented thought, Nietzsche also mounts a critique of contemporary German philosophy that throws into relief and questions its implicit commitment to being contemporary. Philosophy must be freed from its double dependence on the state and the world-historical conception of time that the state fosters.

For Nietzsche, the author of *The World as Will and Representation* is an exemplary philosopher because he stands in stark opposition to D. F. Strauss's Hegel, who links philosophy inextricably to the modern state, a connection Strauss extended to the newly constituted *Reich*. Strauss claimed that the victoriously emergent *Reich* had definitively eliminated the plausibility of pessimistic philosophy. In this respect, consider the parallels with more recent assessments of a similar sort in the culture wars of the last twenty years or so. United States triumphalism following the collapse of the USSR had a similar tone. Intellectuals were advised that their critical stance toward unbridled capitalism had been overtaken by the facts and that the melancholy acceptance of permanent Cold War (marked intellectually by the popularity of "existentialism" and other forms of "irrationalism") had been refuted by history. This response, articulated by Western columnists and pundits, illustrates what Nietzsche discerns as the displacement of thought by journalism, a primary target of the *Unmodern Observations* (and contemporaneous texts such as his lectures *On the Future of Our Cultural Institutions*).

English does not allow us to capture the many ironies that Nietzsche finds in the *Zeitung* (newspaper) or *Zeitschrift* (journal) being taken as the vehicle of the spirit of its time or *Zeitgeist* (perhaps a little French would allow some play between journal and "thought *du jour*"). Another journalistic—or timely, all-too-timely—pronouncement of the culture wars came hard on the heels of the 9/11 attacks. It was said that irony, relativism, postmodernism, deconstruction, and any tendency that could, however vaguely, be related to them—all assumed without much reflection to be forms of pessimism—had been consigned to the rubbish bin of history by the immediate need for a new moral and political seriousness. Much of Nietzsche's Schopenhauer essay is dedicated to freeing thinking from its time, so far as time is understood by journalists or academics who have implicitly adopted the journalistic mentality (for example, Strauss and Hartmann). This will involve contesting the supposed fact that "the new German *Reich* is a decisive and annihilating blow to all 'pessimistic' philosophizing" and undermining "the doctrine, lately preached from all the rooftops, that the state is the highest goal of mankind." As Nietzsche makes clear, the adversarial dimension of genuine thinking is not a complete flight from the timely; rather, "if concerning oneself with one's own age makes any sense at all, then it is a good thing to concern oneself with it as thoroughly as possible, so as to leave absolutely no doubt as to its nature: and it is precisely this that Schopenhauer enables us to do" (*UO* III.4).

Nietzsche's "vivisection of his own time in thought," as he conducted it during the early Bismarck era's culture wars, intensively interrogates the relations among state, universities, and journalism, in which the future of philosophy (or philosophy of the future) is hotly contested. Given changing fashions in academic nomenclature, we might say that Nietzsche is concerned with the state of theory in the expanded post-1970s sense that "theory" has acquired in Anglophone university discourse. What is the point of theory or philosophy and how can it be articulated with a theory of the state? Nietzsche asks whether philosophy (or theory) has veered into journalism (another move prepared by Hegel) by taking its task to be the analysis of current events as they unfold.

Nietzsche's critique of post-Hegelian German philosophy goes to the heart of what links and separates philosophy and journalism as two ways of thinking about time. He charges philosophy with having become journalistic, so far as it aims to analyze and serve its era, *from the standpoint of a certain conception of time*. Underlying the claim that pessimism has been rendered irrelevant by the new Bismarckian configuration of Europe (Strauss), or by the new imperialistic and technological vigor of global capitalism (Hartmann), is the assumption that these are "great events." When it is hastily assumed that we know and understand the "world-historical" importance of such an event (say the *Reichsgründung* or the formation of the European Union), we think journalistically. We lack the perspective of temporal distance. The problem is not only unreflective acceptance of "minor" or "nonevents" as great events, it is the implicit journalistic commitment to the category of the event itself, the event understood as "news" (Nietzsche, as we will see in reading *UO* IV, has his own more radical conception of the event). The *Zeitung* must tell us the news of the day, of the time. When Hegel famously said that the *Zeitung* was the morning Mass of the modern world, he presumably meant something like this: reading the daily paper puts us in touch with higher powers (politics, culture) as well as grounding us in the local (reading local news and business performs a function comparable to the social interchange of churchgoing). Readers of the morning paper (or electronic equivalent) pay their devotions to the world. Beyond that, both participating in the Mass and reading the paper as a modern reader involve accepting a metaphysics. In the case of Christianity this is the mystery of Incarnation with its own account of the beginning and end of time. In the world of the newspaper (or the 24/7 news cycle), it is the perpetual appearance, expectation, analysis, and fear of the event. Events must often be manufactured (most obviously in cultural areas like sports and entertainment, but also in politics) because the *Zeitung* cannot tolerate empty time. The journalist founders when there is no news du jour.

Nietzsche's Schopenhauer gives us the ability to analyze (or deconstruct) our time "as thoroughly as possible, so as to leave absolutely no doubt as to its nature."[23] Schopenhauer gives us an opening into a "thorough" analysis of and confrontation with our time. To be thorough, to go deeply, is to disclose the era's very way of construing time: the journalistic obsession with the event. Thinking that is all too timely takes for granted what it means to live in a time or an era (consider recent expressions like "the new normal," "post-9/11," and the like). Philosophy as generally practiced now, Nietzsche says, the philosophy of the universities, has been captured by the journalistic model, or as Alain Badiou would say, it has been sutured to journalism. Nietzsche calls for a sharp decision in decoupling philosophy and journalism, as Badiou does in demanding a desuturing of philosophy from poetry.[24] Here is Nietzsche exploring the consequences of philosophy's fascination with the *Zeit* of the *Zeitung*:

> Whoever is seeking to answer the question of what the philosopher as educator can mean in our time has to contest this view [that the *Reich* has annihilated pessimism] which is very widespread and is propagated especially in our universities; he must declare it a downright scandal that such nauseating, idolatrous flattery can be rendered to our time by supposedly thinking and honorable men—a proof that one no longer has the slightest notion how different the seriousness of philosophy is from the seriousness of a newspaper. Such men have lost the last remnant not only of a philosophical but also of a religious mode of thinking, and in their place have acquired not even optimism

> but journalism, the spirit and spiritlessness of our day and our daily papers. Every philosophy that believes that the problem of existence is touched on, not to say solved, by a political event [*Ereigniss*] is a joke and ass-backwards philosophy [*Spaass- und Afterphilosophie*] (*UO* III.4).

The political event in question is the foundation of the *Reich*, but Nietzsche's skepticism about such events could easily be extended to those (like the European Economic Community as seen by Kojève or the collapse of the USSR by Fukuyama) that have excited similar responses more recently. "Many states have been founded since the world began," Nietzsche continues, "that is an old story. How should a political innovation suffice to turn men once and for all into contented inhabitants of the earth?" Implicit in the "university philosophy" of Nietzsche's day (and he does not hesitate to name names, almost all now rightly forgotten) is the view that "the state is the highest goal of mankind." A higher duty than serving the state is "destroy[ing] stupidity in every form, and therefore in this form too."

Schopenhauer, so the argument goes, was a true philosopher, and not a mere professor of philosophy. The professors, on Nietzsche's view and in German practice, are civil servants (officials or *Beamten*), so any examination of philosophy's condition must be untimely enough to unearth the connections between state and thought. Accordingly, Nietzsche provides an analysis of the general principles of the cultural and ideological politics of the "so-called nation-state," with emphasis and examples drawn from the contemporary world. Nietzsche gives only a sketchy explanation here as to why the nation-state is only "so-called" (see chapter 3). The sketch in this *Observation* emphasizes the increasingly atomistic and chaotic direction of society. Individuals are less and less bound to one another by natality, ethnicity, and territory. Nietzsche asserts (but does not argue) that his world faces perils of collapse and explosion, and that these dangers are being precariously limited or deferred by the nation-state (Nietzsche was writing not only in the wake of Prussia's wars and the *Reichsgründung* but also during the great world financial crisis and depression of 1873 and in the aftermath of the Paris Commune, which gave him the horrors). Since the natality ingredient in the idea of nation is a factor of decreasing importance, the state maintains itself in two ways, by fear and ideology. As to fear:

> For a century we have been preparing for absolutely fundamental convulsions; and if there have recently been attempts to oppose this deepest of modern inclinations, to collapse or to explode, with the constitutive power of the so-called

nation-state, the latter too will for a long time serve only to augment the universal insecurity and atmosphere of menace (*UO* III.4).

The fear promoted by the state in the early 1870s may at first seem quaint and remote when compared to recent fears of terrorism and weapons of mass destruction. Bismarck, for example, was pursuing a *Kulturkampf* against the Catholic Church and the Jesuits.[25] However, in both cases there is fear of an insidious foreign body, with a secret command center, supposedly striving to undermine or destroy national power. When the Church issued its doctrine of papal infallibility at the same time that the German Reich was proclaimed at Versailles, these apprehensions increased. Nietzsche's response to Bismarckian politics was prescient. Rather than accepting it as the final and destined expression of the age or a people, he understood that it rested on a fragile foundation of chauvinism by stoking the fear of foreign religious agents operating under false pretenses.

The state attempts to develop and maintain citizens' loyalty by promoting an ideology. For example, it underwrites the teaching of philosophy. Nietzsche makes a comparison between medieval and modern solutions to the problem of order. Ever since the Middle Ages, he claims, we (Europeans?) have been struggling with competing forces that threaten to dissolve political and social formations. In those earlier times, it was the Church that held things together through its system of belief, practice, and even a common language (Latin). Now, as we face the threat of "atomistic chaos . . . the state certainly makes an attempt to organize everything anew out of itself and to bind and constrain all those mutually hostile forces; that is to say, it wants men to render it the same idolatry they formerly rendered the church" (*UO* III.4). The state's absolute authority is parasitic on previous claims about the Church's. Here Nietzsche touches again on questions of political theology that will loom larger in his "philosophy of the Antichrist" (announced in *BGE* 256).

We can distinguish two levels in the statist ideology that Nietzsche has in mind. One is specific to the particular state, or even more specifically to the state in a certain era; in modernity, these tend to be forms of nationalism, like the Bismarck era's pan-Germanism. Another level is the sometimes implicit and unspoken idea that it is only through the state that humans are capable of realizing their humanity. Of course, thinkers like Hegel do argue explicitly for this claim. Nietzsche thinks it obvious that the modern state is in the grip of the "money-makers and military despots," foreshadowing what US President Dwight Eisenhower called the military-industrial complex. Such a state must favor thought that promotes its own rule.

Deleuze points to an important Nietzschean contrast between state-oriented and nomadic thought; in *Dialogues* (cowritten with Claire Parnet), Nietzsche is credited with having said everything on this point in *Schopenhauer as Educator*.[26] Deleuze has in mind more than the question of whether philosophers serve the state or think, speak, and write from an independent position. State-philosophy is understood here to involve a certain image of thinking that both parallels a specific political structure and offers legitimacy to the state itself. What the state does is to draw sharp borders between the inside and the outside. One is either within the state or outside it, not only geographically but psychologically or spiritually. The state has a specific identity that characterizes the interiority it establishes. Even when a US citizen travels beyond the state's borders, she carries a passport and retains her identification with the interior. State-oriented thought produces a corresponding mental space of interiority that reinforces and enables the acceptance of the state as a bounded territorial political form. Such philosophy, Deleuze and Parnet say,

> borrows its properly philosophical image from the state as beautiful, substantial or subjective interiority. It invents a properly spiritual State, as an absolute state, which is by no means a dream, since it operates effectively in the mind. Hence the importance of notions such as universality, method, question and answer, judgment, or recognition . . . of always having correct ideas. Hence the importance of themes like those of a republic of spirits, an enquiry of the understanding, a court of reason, a pure "right" of thought, with ministers of the Interior and bureaucrats of pure thought.

Deleuze and Parnet, then, suggest a far-reaching parallel between political and philosophical structures, a convergence of ideology and practice. The political and judicial philosophemes cited are widespread, and deployed sometimes with extravagance (think of Kant on reason's tribunal), so that they do indeed constitute an "image of thought," a picture holding us captive (as Wittgenstein—also a great reader of Schopenhauer—puts it).

Deleuze describes Nietzsche's nomadic "counter-philosophy": "its statements can be conceived as the products of a mobile war-machine and not the utterances of a rational, administrative machine, whose philosophers would be bureaucrats of pure reason" and says that perhaps it is here that Nietzsche announces a "new politics."[27] If the state codifies through law and contract, nomads decodify without recodifying. Deleuze charges that Marx and Freud are ultimately conservative thinkers insofar as they recodify on the basis of a new and perfected socialist state or on a reconstituted story of the family. Kant testifies to the strength of the figure when he dismisses skeptics as nomads; they play a salutary but temporary role in their rebellion against the despotism of the dogmatists, but are fundamentally anarchistic enemies of civilization.[28] Yet not all philosophers follow Kant into the arms of the state (or the expanded version of a multistate pact on offer in *Perpetual Peace*). Nietzsche copied into his notebooks a passage from Ralph Waldo Emerson's essay "History" in praise of "spiritual nomadism": "A man of rude health and flowing spirits has the faculty of rapid domestication, lives in his wagon and roams through all latitudes as easily as a Calmuc [Mongol]. At sea, or in the forest, or in the snow, he sleeps as warm, dines with as good appetite. And associates as happily as beside his own chimneys. Or perhaps his facility is deeper seated, in the increased range of his faculties of observation, which yield him points of interest wherever fresh objects meet his eyes."[29] There is a protophenomenology of different relations to the earth here. History and the history of philosophy belong to the state, geography and geophilosophy to the nomads.

It is not a question of whether the state or its philosophical equivalent comes first (in this sense, it should be noted, neither Deleuze nor Nietzsche holds a crudely reductionist view of the relation between what Marxists call base and superstructure. And, in this respect, recall Marx's declaration, "I am not a Marxist"). Beyond the question of German philosophers' conscious and unconscious acceptance of the ideology of the state that Nietzsche raises in the Schopenhauer essay, there is the question whether philosophers can invent other images of their task. As Deleuze notes, these questions about political and philosophical images are bound up with similar images of the self or psyche. Is the self (or whatever we choose to call it) to be construed as a kind of interior fortress, a ghost in a machine? And even if we abandon such "Cartesian" images, must we also be vigilant about other models of mind and the human that make it a function of language and social interchange? From this perspective, recent forms of critical theory, as in Jürgen Habermas's attempts to outline the structure of an ideal speech situation, are shifts from a monarchical state model to a democratic one, with both presupposing a fundamental parallel of state and thought. Those imbued with the state-oriented model of thought typically suppose any alternatives to the hegemonic model of philosophy must be forms of irrationalism or mysticism. This is to fall into a dualism that simply reinscribes the core image of inside and outside, the reasonably regulated life of the state and the anarchic chaos or undifferentiated unity that lies outside it.

Despite Deleuze's claim that Nietzsche says everything that there is to be said about the two models of thought in *Schopenhauer*, he does not use the term "nomadic" there, although it later becomes a crucial component of his analysis of the hybrid, cosmopolitan, and nomadic multitude (*Menge*). However, in Emerson's *Essays*, one of Nietzsche's inspirations for the *Unmodern* series (especially important, as Stanley Cavell has shown, for *Schopenhauer*), there are passages that constitute a proto-Deleuzian reply to Hegel's understanding of world-history as exclusively a story of states. Early in "History," Emerson gives a significant catalogue of its subject matter: "Epoch after epoch, camp, kingdom, empire, republic, democracy, are merely the application of [the human] manifold spirit to the manifold world."[30] While Emerson speaks the language of "spirit" (like Hegel, but also like Nietzsche in some contexts), spirit expresses itself in the manifold and does not necessarily work toward Hegelian unity. The brief catalogue suggests that, in addition to the Hegelian state, temporary and mobile inhabitations, like the "camp" and what Hegel dismisses as "refuted" state forms (republic, democracy) are equally genuine aspects of "the manifold world." Later in "History," Emerson more explicitly includes the nomadic as a constant dimension of history, not simply its prehistoric presupposition. Since Nietzsche will also develop a related notion of the nomadic later (especially in *BGE*), I cite Emerson in part:

> In the early history of Asia and Africa, Nomadism and Agriculture are the two antagonist facts. The geography of Asia and Africa necessitated a nomadic life. But the nomads were the terror of all those whom the soil, or the advantages of a market, had induced to build towns. Agriculture, therefore, was a religious injunction, because of the perils of the state from nomadism. And in these late and civil countries of England and America, these propensities still fight out the old battle in the nation and in the individual.[31]

Contrary to Hegel, Emerson sees these as two constant rivals, two general, agonistic tendencies whose battle is fought even within the individual, rather than as a sequence of stages. There is no expectation that the state will subsume the nomadic and no exclusion of nomadic peoples from history (we might say that Emerson has dropped the "world" in the Hegelian "world-history"). The need for a religious justification of sedentary life can be read as a deflationary version of Hegel's argument for the necessity of monarchy and state religion, and the application to the contemporary (England and America) foreshadows Nietzsche's claim that nineteenth-century Europe is the incubator of a new nomadism (*BGE* 242).

While Nietzsche does not explicitly develop Emerson's concept of "spiritual nomadism" in the Schopenhauer essay, he does suggest that while modernity's main tendency has been to consolidate the converging models of state, philosophy, and psyche, we can discern other exemplary possibilities in forging images of the human, and each is also a figure of the philosopher. Nietzsche identifies three such responses to modernity, which he associates with the figures of Rousseau, Goethe, and Schopenhauer (*UO* III.4). Deleuze, again inspired by the third *Unmodern*, will call them conceptual personae. Each names a "line of flight" (to borrow another Deleuzian term), that is, a distinctive individual path that rigorously and idiosyncratically commits itself to pursuing a certain way of life. In this case, the three lines of flight are rebellion (Rousseau), cultivation of one's powers (Goethe), and self-sacrifice for the truth (Schopenhauer).

The identification with Schopenhauer involves a highly critical stance toward the state and the danger the state poses to thought that seeks to accommodate it or, worse, enlist in its service. From this perspective, Rousseauian rebellion simply rejects one state form for a future one or for sheer destruction. An idealist of this sort can morph into a "Catilinist," that is, a political nihilist who wants to hasten the destruction of all political institutions. Nihilism is not identical with nomadism. Goethean cultivation, involving "contemplation in the grand style," may preserve and conciliate, but runs the danger of degenerating into philistinism, if its expansive ambitions include some acceptance of modern society and its politics. In this respect, the danger is that the very idea of philosophy will be confused with and displaced by "German dream- and thought-mongering [*Traum und Denkwirtschaft*]" in which thought positions itself for the market (*UO* III.8; interestingly, Marx and Engels use similar language in *The German Ideology* to describe the post-Hegelian German speculation and trade in ideas[32]). The contemporary world is the enemy of "the rebirth of the philosopher" that Schopenhauer heralds. It is "shrouded in humbug [*Flaussen*]; it does not have to be religious dogma, it can also be such bogus concepts as 'progress,' 'universal education,' 'national,' 'modern state,' 'culture war' [*Culturkampf*]" (*UO* III.7). The last term designates Bismarck's *Kulturkampf*, directed against the Jesuits, and more generally the Catholic Church (except for the deviant "old Catholics," who questioned the new doctrine of papal infallibility).[33]

Still avoiding Schopenhauer's (conventionally acknowledged) central texts and ideas, Nietzsche concentrates on the great pessimist's essay "On Philosophy at the Universities." As the famous philosophical story goes, the brash young Schopenhauer offered a course at the same time as the more established Hegel. Not surprisingly, no students appeared, and he did not teach again. Nietzsche aspired unsuccessfully to a chair in philosophy around the time he was writing the Schopenhauer essay. What Nietzsche shares with Schopenhauer is a view of the relations of philosophy, and especially Hegelian philosophy, to the state. Schopenhauer, a man of private means, did not need the economic security of teaching and was famously misanthropic. Given his economic security, Schopenhauer had no need for a teaching position. He recognized that besides wanting a comfortable bourgeois life, a professor (or aspiring professor) might have a need to "shine and show off." Schopenhauer sought the equivalent in literary fame (as with Nietzsche, this was mostly posthumous). That involved creating the character Schopenhauer, lonely and dedicated truth-seeker, who could inspire Nietzsche's encomium. While working on the Schopenhauer essay, Nietzsche was also writing an essay on "Philosophy in the Tragic Age of the Greeks," which explores the question of how the philosopher communicates with his contemporaries by performance and the projection of a persona. Along with his educator, indeed outdoing him in this respect, Nietzsche worked at creating an authorial persona, adapting the strategies and tactics of ancient rhetoric to the print-driven culture of the nineteenth century.[34] The stance of the mysterious, solitary author (evoked at the extreme in *Thus Spoke Zarathustra: A Book for All and None*) is a distinctively recent invention, and possible only in the world of books, print, and mass literacy—that is, precisely in the cultural atmosphere that Nietzsche often denounces.

This construction of an unmodern persona by deploying modern techniques is meant, of course, to offer a compelling figure of the philosopher as an alternative to that of official, university philosophy. Hegel becomes the emblem of that official philosophy through his relation to state and university. Schopenhauer asks "how is anyone who seeks an honest living for himself and his family to devote himself simultaneously to truth, which has at all times been a dangerous companion and everywhere an unwelcome guest?"[35] Take this together with his binary division: "We can divide thinkers into those who think *for themselves* and those who think *for others*."[36] The philosopher as the servant of the university, and therefore of the state, is a comic figure, his supposed commitment to truth shredded by his need to flatter or at least not offend the authorities. Accordingly, Schopenhauer's essay exaggerates the comic: it is a burlesque, carnivalesque screed for the initiated, peppered with Greek and Latin jokes anticipating Nietzsche's polemics, subjecting the now long-forgotten chaired philosophers of his day to withering sarcasm, especially for their devotion to the present in its crudest form. He says of the era that "it calls itself with one of its home-made words, as characteristic as it is euphonious, the 'present time' (*Jetztzeit*); present time indeed, in other words, because one thinks only of the Now and does not venture to glance at the time that will come and condemn."[37]

The projection of the mysterious or flamboyant modern/unmodern philosophical persona became a striking feature of university culture in the 1960s with the emergence of figures like Michel Foucault and Jacques Derrida, both explicitly working in a Nietzschean vein. Each plays a flirtatious game of hide-and-seek with his readers and students, with Foucault notoriously admonishing them not to ask who he is (i.e., not to seek consistency across his writings), while Derrida thematized the "secret" and wrote a number of texts that parody or question the genre of autobiography (*The Post Card*, *Circumfessions*). Given that these thinkers were well paid and privileged French civil servants, they presumably were acutely conscious of the need to demonstrate a sovereign independence of thought and esprit that the butts of Schopenhauer's and Nietzsche's essays lacked. All these maneuvers, by Schopenhauer, Nietzsche, and the "French theorists," contribute to problematizing the Hegelian "we," the monstrous first-person plural Nietzsche had identified in Strauss's Prussian cultural philistines and in the nihilistic self-knowledge of those who emerged in humankind's old age at the end of Hartmann's *Weltprozess*.

Later, in *Ecce Homo*, Nietzsche says that his essays on Schopenhauer and Wagner are both really about himself. In *Schopenhauer* he begins to formulate the question how, outside the university and so outside the state's security, as a thinker of the outside and externality, of earth rather than world, it is possible to find a way of thinking for oneself while developing a mode of addressing an audience, even if one's writing itself is meant in part to conjure that future audience into being. Like the other *Unmoderns*, this project has to do with forms of temporality. The state philosopher is caught in the timely, the time of journalism. At this point in his planned series, then, Nietzsche has undermined three basic ways in which his time has attempted to think time itself. He began with a critique of the self-congratulatory discourse (Strauss's "cultural philistinism") that justifies the present as the historically realized fullness of human possibilities (at least in Germany). The next step was to show that historicism itself involves value-laden choices that can be critically evaluated; more specifically, the tragic and apocalyptic form of end-of-history narratives (like Hartmann's) naïvely reinscribes old stories, while ostensibly combining those "modern" thinkers Hegel, Schopenhauer, and Darwin. The corrective to such grand narratives, whether tragic or comic in tone and structure, is not to surrender to the all-too- timely "now." Each of these approaches to time carries its own political orientation with it. Strauss articulated a conservative political position on the basis of his triumphant "new faith." In accepting the inevitability of the *Weltprozess*, Hartmann found a justification for political passivity and acceptance of what Nietzsche would later call the "last human." The journalistic time of contemporary Germany's state philosophers involves a floating or coasting on the surface of the present, a mode or habitus that puts them de facto in the service of established power. It could be said that Nietzsche's selection of the four focal figures of his critiques betrays a certain limitation to the German "world." Yet he has been challenging that conception of world. In the *Wagner* essay, Nietzsche, however sketchily, begins to formulate a conception of a "great event of the earth" eluding (or twisting free from) the confines of Hegelian world-history. It is the direction of the earth that is at stake. The future is not the necessary conclusion of world-history, but an unpredictable, improbable event in the making that requires our active loyalty.

WAGNER AND THE GREAT EVENT

While Nietzsche had once envisioned the *Unmodern Observations* as a series of thirteen essays, he abandoned the project after the fourth, *Richard Wagner in Bayreuth*. The story of Nietzsche's friendship with the older artist, his enthusiastic heralding of the music of the future (*Zukunftsmusik*, or "music to come") in *The Birth of Tragedy*, private reservations about the master, disappointment with the first Bayreuth festival, and eventual public and critical revision of his own earlier attitude are well known and the subject of endless biographical and psychoanalytical speculation. Here I want to pose these questions: Why did Nietzsche break off the series after the Wagner essay? And how does this paean to Wagner's art cohere with Nietzsche's interrogation of the end-of-history theme and his projection of a possible vigorous unmodern form of philosophy? Perhaps the best indication is given in Nietzsche's later reflections on his obsession with Wagner. In *The Case of Wagner*, the issue is articulated in terms of the struggle with Hegel and Hegelianism, which we have seen as a constant concern of the *Unmoderns*. The late Nietzsche finds the crux of Wagner's work in his claim that his music meant "infinitely more" than mere music, a claim that he was happy to endorse in the last *Unmodern*. This "infinitely more" should be understood genealogically in terms of Wagner's formation: "Let us remember that Wagner was young at the time Hegel and Schelling seduced men's spirits; that he guessed, that he grasped with his very hands the only thing the Germans take seriously—'the idea,' which is to say, something that is obscure, uncertain, full of intimations" (*CW* 10). More specifically, I suggest that the 1876 Wagner essay is fractured by a conflict between Nietzsche's attempt to outline an unmodern, anti-Hegelian conception of history based on notions of event and futurity on the one hand, and his developmental account of Wagner's evolution, which implicitly follows a Hegelian pattern of absorption of the other culminating in a unitary focus. Leaving aside the personal drama, I argue that reading *Richard Wagner in Bayreuth* as an essay in the philosophy of history, in a generally unmarked dialogue with Hegel's corresponding effort, displays both the promise and the partial failure of the series of meditations "under the shadow of swords" that Nietzsche began under Wagner's influence.

The Wagner essay begins by adumbrating the concept of a "great event (*grosse Ereignis*)" and proceeds to consider the question whether it is comprehensible to its contemporaries and what it means for the future. Wagner's initiation of the Bayreuth festival, and in a larger sense his entire artistic career, is taken to exemplify such an event. Hegel too had spoken of "great events," which are for him the turning points of world-history.[38] Hegel's great events are essentially political—more specifically, they are all state-related, having to do with the foundation, transformation, dissolution, and wars of states. In fact, for Hegel, world-history is the history of states: "The state is the divine Idea, as it exists on earth. In this perspective, the state is the precise object of world-history in general."[39] Moreover, for Hegel there is no "historical consciousness" apart from the state, a point he insists on repeatedly. All that precedes the state is prehistory, such as the migrations and wanderings of peoples prior to state formation: "A *Volk* with no state formation (a mere nation/*Nation*) has, strictly speaking, no history—like the *Völker* which existed before the rise of states and others which still exist as wild nations (*als wilde Nationen*)." Without the state there are simply "wild nations" living on the earth; there is as yet no *world* in the strong sense in which Hegel uses this term.[40] Hegel could say of the wild nations what Heidegger said of animals, that they are *weltarm*, world-poor. The life of states is contrasted with the existence of a "people" or "folk" (*Volk*). In the *Unmodern* Wagner essay, the event that its artist-hero ushers in is specifically said to be grounded in and ultimately addressed to the *Volk*, rather than the state (*UO* IV.8). We would expect Nietzsche, following his earlier *Observations*, to develop a non- or anti-Hegelian conception of the great event. As noted earlier, Nietzsche constantly criticizes Hegel and the Hegelians for their commitment to various versions of an "end-of-history" narrative, an end that has either arrived or is on its way. In that view of history, there can be no more great events; for example, in Hegel's theory, once the Germanic world has established the principle that all are free (through the Reformation, the Enlightenment, the French Revolution, and the restoration issuing in the modern state), there is apparently to be no more room for great events. Another version of this idea is Kojève's thought that the battle of Stalingrad, seen as the ultimate conflict of left and right Hegelianism, was the last true event, to be followed by an era of administering the freedom thus attained.

While Nietzsche's own account is not elaborate, a few points stand out. A great event is rare, it is difficult if not impossible to predict or anticipate, and it has a future. "For such an undertaking as that at Bayreuth there were no warning signs, no transitional events, nothing intermediate" (*UO* IV.1); that is, it was not the result of (Hegelian) continuity and mediation. What makes this unexpected event *great* is its transformative power, its throwing the past and the future into a genuinely new perspective. Such events are so rare that Nietzsche offers only two examples, and they form a telling pair. The "last great event," he says, was Alexander's linking of East and West, of Asia and Europe. This involved cutting the Gordian knot that separated the two cultural and geographical spheres and was a generally syncretistic act, mixing together two previously separate domains (*UO* IV.4). In contrast to Hegel's treatment of Alexander (and other "world-historical figures"), Nietzsche does not emphasize the explicitly political side of Alexander's accomplishments here, such as his putting an end to the traditional Greek polis or founding a new imperial form. Instead, it is a cultural achievement that, among other things, opened up the West to Asian religions, eventually including Christianity. Even further from the state-related great events of the Hegelians is the event that Wagner promises to usher in. Wagner had to be more than the possibility of the event, he had to be that event itself. Nietzsche describes him as "the first of the counter-Alexanders," whose task is to unite and focus where Alexander had dispersed, in other words to tie together the threads of European culture in a novel, unified creation.

This contrast of two great events involves an astoundingly high evaluation of Wagner's importance, one running throughout the essay. Wagner is said to have "discovered not only a new art but art itself" (*UO* IV.1), and Nietzsche reminds his readers more than once that in its implications and consequences Wagner's art is much more than art (this is a form of the "infinite meaning" that the later Nietzsche saw as Wagner's Hegelian seduction of his contemporaries). True events, it seems, evoke various forms of respect and adherence. Each phase of Wagner's art attracted its own "circle of acolytes (*Anhängern*)"; if he had attended to them, they would have immobilized his work at that stage (*UO* IV.10). These, we might say, were not worthy of the coming event. Others, disturbed by what they do not comprehend, respond to Wagner with parody and ridicule; these are "facetious newspaper journalists" and the like. Another more authentic relation to the event is found in those whom Nietzsche describes as "we disciples (*Jünger*) of art resurrected" (*UO* IV.1).

The concept of the disciple has been relatively neglected by Nietzsche's readers. Developed and dramatized in later texts (notably *The Gay Science* and *Zarathustra*), the disciple is understood as more than a mere follower or acolyte (*GS* 32, 106; *Z* I.22.3). The genuine disciple grasps a principle—or the meaning of a great event—and embraces it with understanding and unswerving loyalty (*Treue*, a quality that Nietzsche celebrates in Wagner's outstanding characters [*UO* IV.1]). Indeed, the disciple's loyalty to the principle or event may have to take precedence over loyalty to the master who enlightened him (*Jünger* seem to be monotonously male). Discipleship, or putting it a bit more generally, the concept of fidelity, is internal to the event. What gives an event futurity is the fidelity of those who are loyal to it (the *Treue* of the *Jünger*).

Alain Badiou, a contemporary unHegelian and unmodern thinker of the event, offers a much more highly articulated account of similar themes. For Badiou, an event involves a new configuration of meaning that could not have been anticipated on the basis of a previous set of conditions.[41] Events as he theorizes them are rare and have a decisive, universal form. Such events include the French Revolution, which instituted a new conception of human equality. Analogously, the scientific revolution of the seventeenth century transformed the concept of nature, reconceiving all natural occurrences as subject to the same universal laws (as opposed to Aristotelian science, which made fundamental distinctions between the sublunary realm and the rest of nature). These events and others are actualized by those who declare their fidelity to them and open up a future. The universal principles of these political or scientific revolutions become internalized by subjects who become *true* subjects by their fidelity to them, much as in Nietzsche's idea of the disciple. For Badiou, as for Nietzsche, the event in the strong sense opens up a future. The theory and practice of politics and science, although they may abandon some details of the contingent arrangements and views that accompanied the initial event (the form of the French Assembly or some specific theses of Galilean physics) enable and solicit continuing commitments. As with Nietzsche, Badiou's general ontology of the event offers an alternative to Hegelian historical thought with its insistence on the perpetual possibility of the event; on this basis, the end of history is unthinkable, and "great events" need not (but may be) associated with the state. World-history, in Hegel's sense of a dialectically generated progress anchored in the rise and transformations of states, approaching a final culmination, can now be seen as a strategy for evading the event and the future. Toward the essay's end, Nietzsche sums up the contrast: "May sane reason preserve us from the belief that mankind will at any future time attain to a final ideal order of things" (*UO* IV.11).

Yet some differences between Nietzsche and Badiou must be marked. According to Badiou, Nietzsche initiated "the age of the poets," that is, an epoch in which philosophy was subordinated ("sutured") to art. Nietzsche had already declared this move in *The Birth of Tragedy*; his self-inclusion in "we disciples of resurrected art" seems to say something similar. Certainly, Nietzsche's own adoption of poetic and literary voices, and his pride in mastering a vast array of styles, support his being labeled as the first thinker of the age of poets, an age that Badiou sees reaching its apotheosis in Heidegger and terminated by Celan. Nevertheless, while Nietzsche frequently invokes the concepts of event, disciple, and loyalty with respect to art, he is also able to deploy them in a less limited manner. Zarathustra calls successively on his first audience in the marketplace, and then to his disciples, to stay true to the earth and serve its direction; and he describes himself in his noontime mood as resting "now close to the earth, loyal, trusting, waiting, bound to it by the softest threads" (*Z* Preface; I.22; IV.10). In speaking of loyalty to the *Sinn* of the earth, Zarathustra invokes the complex of event, loyalty, and discipleship that were first introduced with respect to art. The event in this case would be nothing less than the elimination of the image of earth as a site of penitence or as ground of Hegel's world-history or Hartmann's world-process. Earth becomes what Nietzsche's hero calls the "human-earth (*Menschen-Erde*)," an earth to be preserved, enhanced, and enjoyed; this requires declaring fidelity to such a direction of the earth. Perhaps this event, one of those creeping up quietly on doves' feet, comes to replace Nietzsche's admiration for Wagner's counter-Alexander "great event" of tying together the various threads of the West into a new, highly focused culture. In *Zarathustra*, the chapter "On Great Events" effectively repudiates Hegel's idea that such events are necessarily state-related, and in suggesting that "the heart of the earth is of gold" offers a different object of loyalty.

More problematic than the somewhat sketchy status of Nietzsche's concept of the great event in the Wagner essay is the fact that it remains in unresolved conflict with the Hegelian dimension of the text. Not only is there a deeply Hegelian side of Wagner, as Nietzsche later recognized, but his own work of 1876 is infected by it, and this leads to an incoherence, which helps to explain why the series of *Observations* could go no further. What Nietzsche required was a non-Hegelian concept of history and an alternative notion of the great event; but it was his very subject, Wagner, who both tempted him in this direction and made the task impossible at that point.

There are then two distinct and inconsistent lines of thought or two voices in the last of the *Unmodern Observations*. We have traced one of these and now we turn to the other. This aspect of the essay is a story of how Wagner "became what he is," a narrative of how he became "whole and wholly himself" (*UO* IV.1, 3). Nietzsche realizes that he must explain how Wagner emerged from his chaotic past—youthful revolutionary enthusiasm, experimentation with various musical forms, polemical writings, and flamboyant public persona—to become the focused artist of Bayreuth. In this interpretation, Wagner evolves and purifies himself, a process mirroring the development of his most significant characters. Wagner is said to have a super-German talent for learning and absorbing a staggering variety of thoughts and styles (*UO* IV.3). While at first this sounds as if he shares the German weakness for history that Nietzsche had diagnosed in the second *Unmodern*, he is also described as having an equally unusual talent for focusing and integrating, a talent thrown into relief by a contrast with Goethe, whose power is more dispersive, "like a many-branched river system which fails to sustain its full force as far as the sea but loses and scatters at least as much on its windings and meanderings as it bears on its estuary." By purifying and redeeming himself, Wagner makes possible the purification and redemption of the disciples of art. On the one hand he is seen as embodying an event that shows the ineluctable futurity of life and art; on the other he is sometimes described as bringing art to fulfillment, a description reminiscent of Hegel's notorious claim that the idea of art is completed and realized in his present (a view sometimes described as the death of art).

This way of understanding Wagner's development oddly resembles Hegel's account of the growth and completion of the Germanic world in his world-history. As we saw earlier, Hegel's account of the Germanic peoples as they emerge into history hinges on their enormous capacity for absorbing and transforming what was initially foreign, notably Roman culture and Christianity. Unlike the other world-historical peoples, who begin by defining themselves and then encounter the other, "the Germans, on the contrary, began with self-diffusion—deluging the world, and overpowering in their course the inwardly rotten, hollow political fabrics of the civilized nations. Only then did their *development* begin, kindled by a foreign culture, a foreign religion, polity and legislation."[42] The development proceeds through the establishment of the Holy Roman Empire, the Reformation, and finally the modern state, with its realization of the ultimate unity of the secular and the spiritual. The freedom and loyalty that Hegel takes to be characteristic virtues of the Germans are very close to what Nietzsche discovers in Wagner, and which he also attributes to the German spirit. Loyalty or fidelity (*Treue)* is ambiguous in the Wagner essay, sometimes lending itself to the Hegelian narrative, and at others to Nietzsche's sketchy alternative conception of fidelity to great events.

It has been known for some time that Nietzsche's notebooks around 1874, when he was planning and beginning the Wagner essay, contain much more critical remarks on the ideal figure of the published work. In these private observations, Nietzsche sees Wagner's music, his expectations of his audience, and his conception of state and society as distorted by an extreme histrionic talent and ambition, a theme developed explicitly in such later writings as *The Case of Wagner* and parodied in the figure of the Enchanter (*Zauberer*) in *Zarathustra*.[43] Whatever doubled consciousness was necessary for Nietzsche to hold both perspectives simultaneously, what he apparently failed to see in the 1870s was the depth of Wagner's Hegelianism. He was unable then to acknowledge his own reliance on typical Hegelian themes in producing a portrait of Wagner's evolution. It is an evasion that becomes intelligible when we realize that Nietzsche's unmodern thinking seemed to him to require an exemplar, a figure who could demonstrate once again the possibility of great events. Since the last great event, in his view, was Alexander's fusion of East and West, the need was extreme, so the denial can be understood psychologically, if not philosophically. Ten years later, in *Beyond Good and Evil*, Wagner and Hegel are said to be similarly German in their cloudiness and indeterminacy; Wagner especially is classified as a European hybrid. The artist whose name Nietzsche had earlier taken as the sign of a nonstate, non-Hegelian, deeply Germanic-European great event then becomes one symptom among others of Europe's hybrid, nomadic, and diverse culture of the multitude (*Menge*).

CHAPTER 3

Living on the Earth: States, Nomads, Multitude

> The state is the divine Idea as it exists on earth. In this sense the state is the precise object of world history.
>
> HEGEL, *Introduction to the Philosophy of [World] History*

> All significant concepts of the modern theory of the state are secularized theological concepts.
>
> CARL SCHMITT, *Political Theology*

> State I call it where all are poison-drinkers, the good and the base; state, where all can lose themselves, the good and the base; state, where the slow suicide of all is called—"life."
>
> ZARATHUSTRA "ON THE NEW IDOL" IN NIETZSCHE, *Thus Spoke Zarathustra*

The previous chapter followed Nietzsche's assault in his *Unmodern Observations* on a complex of related modern, all-too-timely ideas. He saw these ideas as blocking the sense of futurity crucial to posing the question of earth's direction. The *Unmodern* series broke off with his sketchy and flawed attempt to outline the form and content of the "great event" that was to bear the name Wagner. In his next publication, Nietzsche began to subject the structure of the national state to a more structural form of analysis, moving beyond his condemnation of its contributions to cultural philistinism and its smug, self-congratulatory Whig history. *Human, All-Too-Human* marks the beginning of this analysis as Nietzsche starts to question the state's claim to legitimacy, the viability of its nationalist ideology, its problems in dealing with mobile populations, porous borders, and its reactive security hysteria. While this chapter begins by examining some texts from *HAH*, it will follow Nietzsche's development of these themes more thematically than chronologically. The state, as Zarathustra says, is "the new idol." Like the religious idols of old—and attempting to legitimize itself by analogous ontological claims—the state aims at usurping power over the earth, transforming it and its peoples into a manageable world.

STATE OF EXCEPTION: THE LOGIC OF THE "DECIDER" AND THE DECLINE OF THE STATE

Since the United States' commitment to a "global war on terror" following the attacks of September 11, 2001, there has been increasing attention in Anglophone political thought to the idea of the "state of exception," the political mode/act by which the state, more specifically the sovereign power, suspends some of the state's own laws, including even parts of its constitution, in order to deal with such emergencies as foreign or civil war. Now political thinkers are all aware of the maxim of Carl Schmitt, the mercurial jurist notorious for his involvement with the Nazi regime: "Sovereign is he who decides upon the exception."[1] There are echoes of the principle in George W. Bush's self-description as "the decider," a description that seems to have been stimulated by members of his legal staff. Nietzsche was not only aware of this kind of move in contemporary Europe, but also attuned to the analogy Schmitt was later to make explicit between political and theological sovereignty. For Schmitt, the sovereign's authority in and over the state is comparable to God's rulership of the world; suspensions of the constitution are parallel to divine miracles that suspend the laws of nature.

Nietzsche's most obvious reference to the theme is in section 475 of his 1878 book *Human, All-Too-Human* (the date is significant, as we'll see, because of the contemporary practices of the German *Reich*). The reference has gone unnoticed, since the aphorism is typically read in order to illustrate or explore Nietzsche's thoughts about Jews and anti-Semitism, with readers commenting on the complex attitudes in his observations.[2] For example, from a biographical and psychological standpoint, the text has been understood as marking Nietzsche's turning away from Wagner's anti-Semitism. When Cosima Wagner read the book, she said that it was the voice of Israel that now spoke through Nietzsche, charging that "Israel had taken over" through the malign influence of Paul Rée, his Jewish friend.[3] On the one hand, Nietzsche speaks of modern Jews as possessing "energy and higher intelligence" accumulated through a long school of suffering. And if he allows that "perhaps the youthful stock-exchange Jew is the most repulsive invention of the human race," he contextualizes this by saying that "every nation, every man possesses unpleasant, indeed dangerous qualities: it is cruel to demand that the Jew should constitute an exception." He asks how much must be forgiven a people who "not without us all being to blame, have had the most grief-laden history of any people and whom we have to thank for the noblest human being (Christ), the purest sage (Spinoza), the mightiest book and the most efficacious moral code in the world." If the contemporary reader wants to see the case for decoupling Nietzsche from his popular associations with Nazism and other forms of anti-Semitism, or explore the complexity of his attitude toward Jews, this aphorism is a good place to start.

Yet what is the explicit subject of this text, which bears the title "The European human (*Mensch*) and the abolition of nations"? We might note that all of Nietzsche's comments about the Jews here are marked as consequences of his initial argument concerning the fragility of the modern state. Like his contemporary Marx, but for rather different reasons, he imagines that the state will eventually disappear. The second section of the aphorism is marked off from what precedes it by a dash, one of Nietzsche's characteristic stylistic flourishes: "—Incidentally (*Beiläufig*): the problem of the Jews exists only within national states . . . ." "The Jewish question" (as Marx calls it) is something that arises only in the nation-state, revealing the state's conceptual incoherence and fragile stability.

What then was Nietzsche thinking about the problem of the national state? Why is the "problem" of the Jews framed as an illustration of his analysis? Readers have not always attended to the aphorism's title or to the fact that it is embedded in the chapter called "A Glance at the State." They may rush past the first section, which develops this theme, and on to the second, which Nietzsche introduces as a consequence of its main thesis. What is that thesis? Nietzsche rapidly sketches a case that the increasingly common cultural and economic life of Europeans, and the growth of "nomadic life among all those who do not own land"—mobility—"are necessarily bringing with them a weakening and finally an abolition of nations," which must eventually produce a mixed race of European humanity.[4] Along these same lines, he suggests (apparently for the first time) that "one should not be afraid to proclaim oneself simply a good European and actively work for the amalgamation (*Verschmelzung*) of nations." In this context he remarks on the Germans' special ability to mediate among peoples. A twenty-first century reader should be struck by Nietzsche's analysis of the obstacle in the way of this process: "This goal is at present being worked against, consciously or unconsciously, by the separation of nations through the production of *national* hostilities . . . this artificial nationalism is in any case as perilous as artificial Catholicism used to be, for it is in its essence a forcibly imposed state of exception and siege (*gewaltsamer*

Noth- und Belagerungszustand) imposed on the many by the few and requires cunning, force, and falsehood to maintain a front of respectability."[5]

The argument, then, proceeds through several stages. First, Nietzsche claims something about Europe. Despite the time's noisy nationalism and chauvinism, he discerns a real counter-movement, in which Europeans are becoming increasingly mobile or "nomadic." This leads to a loosening of traditional national ties and identities. Here Nietzsche is again effectively repudiating Hegel's world-history, which dismisses the exclusion of the wanderings and migrations of peoples as nonhistorical, and assigns a national character to each state. Nietzsche takes nomadism to be an indisputable facet of European modernity. It is not a temporary disruption or aberration. Here we can compare Nietzsche's claim with Hegel's by considering how Hegel marginalizes two significant geopolitical phenomena, the contemporary rise of the United States and the seven centuries or so of powerful Islamic states, empires, and caliphates. Hegel sets up a logical contrast between two more or less contemporaneous developments, the wanderings of the Germanic *Völker* and the spread of Islam. The *Völker* are merely *particular* in their origins, tied to arbitrary, contingent events and traditions; in opposition, Islam is the rule of *abstract universality* and is especially suited to the Arabs, who roam the wide expanses of the desert, which Hegel compares to the boundless sea. In the Islamic realm, Hegel sees nothing but an episodic succession of wars, caliphates, and kingdoms where "nothing firm abides."[6] The moment of *individuality* comes with the empire of Charlemagne, uniting a number of Germanic tribes, drawing a firm line with Islam, and instituting the outlines of a state. While Hegel did not claim to predict specific futures, he certainly intended to exclude certain possibilities. He denies that the United States in its democratic, secular form, and Islam as a religious-political phenomenon, can be genuine players in the field of world-history. This highlights the importance of Nietzsche's implicit rejection of the original exclusion of the nomadic, which generates not only Hegel's version of "so-called world-history," but also all of its variants (Strauss, Hartmann, and more recent narratives), which fail to recognize the instability and limits of the national sovereign state. In this respect, Hegel and his heirs are still in thrall to the principles of national sovereignty laid out in the 1648 Peace of Westphalia.

For Nietzsche, on the contrary, insofar as the nation-state supposes a population of common ethnic origins and culture, it finds itself in an intrinsically problematic position, because mobility and mingling contribute to the formation of a "mixed race," a process he seems to welcome (or at a minimum accepts with equanimity). The second stage of Nietzsche's argument, then, is that there is no point in resisting this inevitable process, which may be an important experiment in the European form of life, perhaps with trans-European implications. While some of the migration and mobility that Nietzsche alludes to may have to do with individuals seeking employment, opportunity, or freedom from old, restrictive traditions, he also seems to be thinking about the movements of families, subcultures, and groups. In his vocabulary (coinciding generally here with that of geographers and historians), the nomadic designates a collective rather than an individual mode of inhabiting the earth. Finally, in this first part of the aphorism, Nietzsche notes what retards the transformation or abolition of the national state. It is the state's scare tactics, its exaggeration or fabrication of external or internal threats to the population's security.

The state uses scare tactics to justify what Nietzsche calls a "*Noth- und Belagerungszustand*" or state of exception and siege. These technical terms of German law are related to other legal forms in various nineteenth-century states. Politically, the state of exception (*Ausnahmezustand* is the most frequent German term) is the assumption by a sovereign power of the right to suspend laws and constitutions, as in the United States version, during the Bush 43 administration, of the "unitary executive." The state of exception is imposed by a single executive or executive group that declares itself to be the sole "decider," and its signature decisions, Nietzsche says here, are those geared to the fomenting of national hostilities or fear of internal subversion in order to maintain the artificial unity of the state. This is formalized in Schmitt's famous definition of sovereignty: "Sovereign is the one who decides on the exception."[7] In this view, sovereign power manifests itself in the ability to limit or suspend the application of laws and constitution. Since any state may be beset by an emergency, and since the time and character of the emergency are necessarily unpredictable, the state is constantly haunted by the possible declaration of the state of exception. Even during "normal" times, with no suspension of constitution or law, it is the sovereign who must judge that no emergency in fact obtains. The sovereign is still the one who *could* or *would* declare a state of exception. The sovereign, then, is the ultimate "decider" in the political sphere, a title that George W. Bush seems to have arrogated to himself on the advice of his neo-Schmittian staff.[8]

The most notorious provision for a state of exception was article 48 in the constitution of Weimar Germany, which empowered the president to suspend law in time of emergency. Shortly after taking power, the Nazis invoked article 48 and proceeded to do away with a series of rights and liberties, although the constitution itself was never formally revoked. Giorgio Agamben analyzes the history and contemporary importance of these acts on the part of the sovereign in *State of Exception.*[9] Since 2001, in the United States the executive branch has both openly and secretly introduced exceptions to the Constitution and other laws, such as those governing the surveillance of citizens and the right of habeas corpus, in the name of the emergency that goes by the name of a "global war on terror." As Nietzsche puts it, and as the record of the Bush II administration confirms, this "violently (*gewaltsamer*) imposed state of exception and siege is inflicted on the many by the few and requires cunning, lies, and violence (*List, Lüge, und Gewalt*) to maintain a front of respectability." Whether cooking the intelligence on Iraq's supposed weapons of mass destruction or retaliating against truth-tellers and whistle-blowers (as in the Valerie Plame scandal), the executive pays lip service to the need for a rational justification of policy. Nietzsche notes that the state of exception does not serve "the many (the peoples [*Völker*])" whom it ostensibly protects, but "the interests of certain princely dynasties and certain classes of business and society."

Might Nietzsche help us in understanding the passions and the logic at work here? He is writing in 1877–78, just six years after the Franco-Prussian War and Bismarck's founding of the German *Reich*. Nietzsche distanced himself from what he saw as Germany's arrogant capitalization on its victory almost immediately. As the aphorism on the decline of the nation shows, Nietzsche's critique has both political and cultural dimensions. Nietzsche would have been familiar with article 68 of Bismarck's constitution, the ancestor of the notorious article 48 in Weimar's.[10] Throughout the later nineteenth century, various European and American states enforced states of exception and suspended constitutional guarantees; Lincoln, for example, suspended the right of habeas corpus during the American Civil War. The French Constituent Assembly declared a state of siege following the fall of the July monarchy in 1848. Italy, as Agamben says, "functioned as a true and proper juridico-political laboratory for organizing the process" by which what initially was a disdained and exceptional instrument became a normal means for the production of law, with Palermo, Naples, and Sicily declaring frequent states of exception.[11] Of course, it can be argued that many of these declarations were indeed necessary in the circumstances. However, what Agamben notes is what Nietzsche claimed: the general tendency was to normalize the state of exception itself as a way of strengthening state power. The state of exception or emergency (*Nothzustand* or *Ausnahmezustand*) is thought of as a response to a genuine and pressing need (*Nothstand*). In the same chapter on the state, Nietzsche observes that officials have a vested interest in exaggerating emergencies; even if the knowledgeable among the population are aware of the exaggeration, the more ignorant majority are more susceptible to being swept up in the "investigations, punishments, undertakings, reorganizations" that follow upon the exaggerated emergencies (*HAH* 448; cf. *D* 179). To see the contemporary parallels, think only about the fear promoted by the second Bush administration when it frightened the US population with the specter of Iraqi atomic weapons (the "smoking gun" that could be a "mushroom cloud"). Indeed, so powerful were the forces of exaggeration that at the time of the US invasion of Iraq, fully 45 percent of the US population believed that Saddam Hussein's Iraq was a major force behind Al Qaeda's 9/11 attack on their "homeland." (The "homeland" is a term of art in the recent intensification of the security state; as we shall see, Nietzsche's analysis of "the new idol" and of "peoples and fatherlands" offers resources for deconstructing this rhetorical maneuver.)

The typical justification offered for such a state of exception, Nietzsche says later in *Human All-Too-Human*, commenting on the Inquisition, is the claim of exclusive truth and virtue, which must be preserved for the sake of all mankind (*HAH* 633). As the Catholic Church justified the Inquisition by appeal to faith and authority, so the US suspension of liberties in support of the "war on terror" appeals to such religious principles as "God's gift of freedom," even if the war involves torture and abuse reminiscent of the horrors of the Inquisition (e.g., waterboarding). In a later section of *HAH*, Nietzsche gives a succinct account of how kings and emperors—the "decider" figures of his day—are tempted to avoid marginalization by the growing democratic and constitutional forces of the time. Rather than becoming "splendid ornaments," they "cling with their teeth to their dignity as *warlords*: for this they require wars, that is to say states of exception (*Ausnahmezustände*) in which that slow constitutional pressure of the forces of democracy lets up" (*WS* 281).

From a Nietzschean perspective, Schmitt's understanding of sovereignty as a theological concept is not surprising. Following the principle that the exception is more important and revelatory than the norm, Schmitt sees God and the sovereign as typologically analogous ultimate instances of decision, power, and authority. Just as the idea of a unified world is impossible without God, Schmitt suggests, the unified state requires a sovereign with an unhindered power of decision. Nietzsche could very well accept this parallel, but would reject both God and the sovereign as understood by Schmitt. In Nietzsche's crucial aphorism on the death of God (*GS* 125), the "madman" who announces this in the marketplace says that the import of the news will dawn only very slowly, as humans gradually grasp what it means to live without an absolutely centering concept. For many centuries, he suggests, we will still be living with God's shadow, that is, with various substitutes for the absolute and centering instance (*GS* 108). In this light, Schmitt is a shadow artist, a theorist who explicitly attempts to endow the state with the attributes of the divinity. As Nietzsche had his madman ask why we don't yet smell the odors of God's decomposition, so his analysis of the slow but inevitable decay of the national state points to a parallel decentering in the political sphere. To sum this up: "The belief in a divine order in the realm of politics, in a sacred mystery in the existence of the state, is of religious origin: if religion disappears the state will unavoidably lose its ancient Isis veil and cease to excite reverence" (*HAH* 472).[12] Later, in *The Antichrist*, Nietzsche will charge Christianity with creating the very states of emergency (*Notzustände*) to which it appeals to justify its parasitic tyranny (*AC* 62). As we will see, such analyses are part of his program of turning the themes of Christian political theology against itself.

At the same time, Nietzsche rejects an activist anarchism that attempts to accelerate the state's decay. In the aphorism last cited on "Religion and government" from "A Glance at the State," he produces a schematic genealogy of their relationship. It is a story of the progressive and systematically linked decay of both, culminating in the thought that "a later generation will see the state too shrink to insignificance in various parts of the earth" (*HAH* 472). Note that Nietzsche speaks of the earth rather than the world as the larger horizon here. The state or government (*Regierung*) has an obvious interest in allying with religion insofar as it wishes the populace to "patiently submit to instructions from above (in which concept divine and human government are usually fused.)" Yet this use of religion will eventually give rise to a certain enlightenment, the beginning of "freespiritedness" among the rulers, from which, Nietzsche leaves us to infer, only more skepticism can develop. (We might see here the roots of Leo Strauss's meditations on religion, state, and philosophy.) If government happens to develop in a democratic direction, the plurality of religions will lead to the state regarding religion as a private affair of conscience. In this atmosphere, sects will flourish freely, the more enlightened will become at least privately irreligious, and many of those still moved by religion will become hostile to the state, while in reaction another party will become increasingly enthusiastic supporters of the state. The emptiness that growing secularism has left in their hearts will be filled by this new devotion. So far, Nietzsche's genealogy could help to explain the rise of fundamentalist and antifundamentalist politics in an avowedly secular state like the United States. Yet after such "transitional struggles," which may last a long time, "the attitude of veneration and piety" toward the state will be undermined, and it will begin to be seen in a pragmatic and utilitarian perspective. At this point much of the work of government will be reassigned to "private contractors"—"outsourcing" is the current word—another sign of the impending "decline and *death of the state*." Although Nietzsche does not mention Hegel here, this would surely entail the collapse of his state-centered world-historical narrative; on the post-state earth, "a new page will be turned in the storybook of humanity in which there will be many strange tales to read and perhaps some of them good ones." Just as the domination of the organizing principle of the racial clan gave way to the family and then to the state, so humanity will eventually hit upon "an invention more suited to their purpose than the state." (Nietzsche continues to speak of the earth as the sphere of human activity, eschewing the vocabulary of "world" and "so-called world-history.") However, there is no point in working actively for the "dissemination and realization" of the idea of the state's dissolution, since even the most far-sighted can have little sense of the consequences this would entail. On what may seem a surprisingly conservative note, Nietzsche concludes: "Let us therefore put our trust in 'the prudence and self-interest of humans' to preserve the existence of the state for some time yet and to repulse the destructive experiments of the precipitate and the over-zealous!" Now it is clearer why Nietzsche calls this section of *Human, All-Too-Human* "A *Glance* at the State." Rather than being the basis of world-history and the foundation of art, religion, and philosophy, as for Hegel, the state is simply the most recent organizational form of human life on earth. Its hypertrophic development in the era of nationalism, Nietzsche says, is temporary in the larger scheme of things. A glance (*Blick*), in Nietzsche's vocabulary, is a quick but comprehensive look that delineates structure, meaning, and context. Such a glance is sufficient to put the state in its place.

Nietzsche extends his account of the state of exception by showing how its logic is already at work in our internal psychic economy. This picture of the self as a community is comparable to Plato's reading justice or injustice in the state as modes of the individual human soul writ large, and Freud's understanding of the self as a complex balance of id, ego, and superego. In a chapter on "The Human Alone with Itself" he writes:

> *Self-observation.*—The human (*Mensch*) is very well defended against himself, against being reconnoitered and placed under siege (*Belagerung*) by himself, he is usually able to perceive of himself only his outer walls. The actual fortress is inaccessible, even invisible to him, unless his friends and enemies play the traitor and conduct him in by a secret path (*HAH* 491).

The human being is in a state of exception with regard to himself, a strange doubling logic, but no stranger than the explicitly political state of exception. We do not know our own well-defended inner fortress. We resist self-knowledge by declaring a state of exception that makes our core as uncanny as the sovereign imposition of a law that suspends law. Both in the case of the individual and in that of the state, friends and enemies, forces of the *outside*, are necessary to break the defenses and the martial law that claim justification to resist the siege. The tightly constructed individual identity (think of Freud's superego) and the nationalistic political state are powerful yet fragile, and the struggles to dissolve both of these artificial unities have a common structure. Many individuals borrow the principle of the state of exception from the political realm to make their own injustices acceptable: "There are not a few who understand the unclean art of self-duping by means of which every unjust act they perform is reminted into an injustice done to them by others and the law of exception for necessary defense (*Ausnahmerecht der Notwehr*) is reserved to what they themselves have done" (*AOM* 52). Such people are in need of an "intervention," in contemporary therapeutic language.

Perhaps most unsettling, given Nietzsche's project of opening up possibilities for thought that are free of state-oriented concepts (as in *Schopenhauer as Educator*), is the danger that philosophers themselves may begin to *think like states*. Nietzsche argued this in the Schopenhauer essay with respect to the Hegelian and journalistic state-employed philosophy professors of Bismarck's Germany. Yet the problem is more deeply rooted. In a long aphorism of *Dawn* he takes a critical look at aging philosophers, once original and daring thinkers, who end by wanting to institutionalize their ideas. This can stifle the creativity in others that they had enjoyed themselves. Toward the end of their careers, thinkers like Plato and August Comte become weary, shift their concern from fresh thinking to preserving their legacy, and seek to enshrine their ideas in a "temple of enduring stone." Probably the "most dangerous characteristic" of this weariness is their "belief in their own genius, which usually assails great and semi-great men of the spirit only at this frontier of their life: the belief that they occupy an exceptional position (*Ausnahmestellung*) and enjoy the law of exception (*Ausnahmerecht*)" (*D* 542). When the aging philosopher comes to believe that the preservation of his thoughts is threatened by misunderstanding or rivals, he appoints himself as sovereign (or "decider") in the realm of the spirit. Nietzsche wrote about a legend according to which Plato took pains to have all of Democritus's writings destroyed; more recently Comte attempted to establish a church of positivism.[13] The thought of the state privileges interiority and the maintenance of its borders. In a note of 1874, Nietzsche writes that philosophy "is gradually turning into nothing but the guarding of borders."[14] Remarks along these lines will be familiar to those Anglophone teachers and students of philosophy who lament the tendency of "analytic" philosophers to exclude other kinds of thinkers from the charmed circle of "real" philosophy. This is known colloquially in the United States as "circling the wagons," which suggests the territoriality of these border conflicts and some typical defense strategies.[15]

Nietzsche's frequent denunciations of anarchism should be read in the light of his diagnosis of the decline and death of the state. The nineteenth-century anarchist movement that Nietzsche rejected took two main forms. From his perspective, utopian anarchists like Kropotkin were naively deluded in thinking that there could be a quick and relatively painless transition to a world of mutual cooperation free of the state. Revolutionary anarchists like those inspired by Bakunin (Wagner's ally in 1848) were nihilists whose violence and terrorism were fueled by a powerful, raging *ressentiment*; their destructive activities could only encourage the state to declare a state of exception. Like Marx, Nietzsche saw the future of the state and its "withering away" as embedded in a complex historical movement, although they had quite different takes on what that process is.

FROM WORLD-HISTORY TO THE HUMAN-EARTH: ZARATHUSTRA'S GEOPOLITICS AND GEOAESTHETICS

Nietzsche himself was one of those mobile Europeans whose lifestyle, he thought, foreshadowed the decline of the national state. He had been a stateless person since taking up his professorship at Basel in 1869. Frequently on the move, he followed the seasons, living for a few months at a time in Sils Maria, Venice, Genoa, Turin, or elsewhere. Nietzsche's life was defined in many ways by the trans-European rail and postal systems, as he corresponded with friends, family, colleagues, and publishers throughout an itinerant life. While many have written in a psychologizing vein about the solitude evidenced by Nietzsche's travels, we could see his wanderings more productively as a way of inhabiting and discovering the earth. Nietzsche's mobile personal geography alerted him to the globalization of space and time through the consolidation of a striated perspective on the earth. He lived mindfully in a time increasingly dominated by techniques of measurement, transportation, and control of speeds and flows. The poetic and prophetic tone of *Thus Spoke Zarathustra* resonates with the transformation of space and time in nineteenth-century Europe. At the same time, he read the latest works on human and physical geography, situating him like no other philosopher of his time with respect to the changing nature of humans' relation to the earth, and adding depth to his concept of the *Menschen-Erde.*

Speaking in the marketplace, in the place of exchange, his first entry into the public sphere, Zarathustra denounces the overly measured world of the last human, who could be the product and final witness of Hartmann's *Weltprozess*: "The earth has become small, and on it hops the last human, who makes everything small. His race is as ineradicable as the earth-flea; the last human lives longest" (*Z* I "Prologue" 5).

In the late nineteenth century, the world was becoming markedly smaller through a market-driven excess of measurement. Nietzsche was perhaps the first railway philosopher; not only did he live an itinerant, nomadic life traversing the European continent and corresponding through the bureaucratized postal system, but his thought also responds to the globalization spearheaded by the railways, telegraph, and telephone. In his working notes for *Dawn*, Nietzsche strategizes how to present aphorisms to the modern, mobile, traveling reader. The reader who hits upon paragraph 454 of the published version will be told that such books are not meant to be read straight through but are best used as episodic provocations while traveling (today you could subscribe to a Nietzschean aphorism of the day on your mobile device).[16]

The primacy of the earth, the question of the state, the possibility of great events are all dramatized in *Zarathustra*. I focus on these topics in a brief conspectus that necessarily minimizes the poetic and rhetorical complexity of this unclassifiable "book for everyone and no one." The alternative that Nietzsche's Zarathustra proposes to the shrinking, globalized world of his time is called loyalty to the earth, not world-history or world-process. Zarathustra challenges his listeners to ask: "What shall be the direction [or "meaning," *Sinn*] of the earth?," urges them to "remain true to the earth," to think with an "earthly head" (*Erden-Kopf* [*Z* I "On Believers in a World Beyond"]), and create a truly human-earth (*Menschen-Erde*, [*Z* I "On the Bestowing Virtue" 2; *Z* III "The Convalescent" 2]). Robert Gooding-Williams says that "earth" here is a metaphor for the human body and its passions.[17] Certainly, Nietzsche is concerned with what it means for human life to flourish in a thoroughly immanent world. But this, I suggest, involves more than individuated human bodies; Nietzsche could have agreed with Marx, who called the earth the human's inorganic body.

In the course of his work, Nietzsche's horizon expands from a focus on Greece and Germany, to a European perspective; he eventually says that even an understanding of Europe requires a "trans-European eye."[18] I suspect that reading Nietzsche in a Heideggerian mode, through the prism of Hölderlin's

Greek and German earth, risks what Foucault called the return and retreat of the origin, analogous to the nostalgia and site fetishism that mar Heidegger's thought. In contrast, a feminist phenomenological and psychological critique of Nietzsche, like Luce Irigaray's in *Marine Lover*, challenges us to ask what is involved in Nietzsche's complex play of elemental and geographical tropes.[19] Irigaray charges that Nietzsche valorizes the air, mountain heights, and flight, consequently obscuring and marginalizing the fluidity and depths of the sea. While freeing Nietzsche from the burden of history, Irigaray makes his language of the earth into one more expression of patriarchal essentialism, missing the fluidity that he sees in the "elements" of the earth, and implicitly rejecting his experimentalism. Both approaches are oddly reminiscent of Ernst Bertram's *Nietzsche: Attempt at a Mythology*.[20] Bertram's World War I-era book attempts the construction of a new Germanic mythology, extracted diagonally from Nietzsche's texts, concentrating on historical figures and geographical sites, such as Venice and Portofino. Bertram mines Nietzsche's writings to construct a Germanic, mythological escape from modernity and all but ignores his theme of the "good European's" emergence in a globalizing world endangered by the "insanity of nationalism," maintained by proclamations of states of exception enforced by deception and fear-mongering (*HAH* 472–75).

Nietzsche adopted the external form of Hartmann's and globalization theory's question about the direction of the earth, and reframed it, asking how humans (or posthumans) can flourish on the earth. Consider the two books that he said expressed the two sides of his general philosophical project, *Thus Spoke Zarathustra*, the yes-saying version, and *Beyond Good and Evil*, the no-saying one. Superficially, they are in sharp contrast. Myth, drama, song, and landscape seem very different from analysis, argument, and a detailed focus on the follies of contemporary Europe. At a deeper level, they are indeed the yin and yang of a philosophical vision of human life on the earth and its prospects for the future.

Many philosophers and critics have taken on the challenge of explicating *Zarathustra*'s song of the earth. At what may be the philosophical crux of *Zarathustra*, when the hero literally wrestles with the thought of eternal recurrence, he agrees with his talking animals not on their interpretation of that teaching (which he dismisses as a "hurdy-gurdy song") but on their statement that the world awaits him "as a garden." One of the many candidate genres for this unclassifiable book for all and none is that of the philosophical landscape poem provoking us to rethink our relation to the earth. This is not the place to explore those eco-readings in depth or to attempt another such comprehensive analysis.[21] Instead, I will presuppose the outline of the work's aesthetico-political geography and suggest a reading of two rather explicitly political sections, one on the state, the other on great events. Each embodies the critical relation to Hegelian world-history ingredient in Nietzsche's conception of the human-earth. Descending from his mountain cave, Zarathustra begins an adventure that Nietzsche marks in the four parts of the text. The town of the Motley Cow, where he teaches in Part I, is a generic city; scholars have noted that its name may allude to Plato's description of democratic Athens or to a site of Buddha's mission. In any case the emphasis seems to be on an urban environment where many ideas are in the air and young people are looking for guidance.

In a talk in or near this town, Zarathustra issues a prophetic denunciation of the state, a discourse "On the New Idol." With its Biblical resonances and parodies, we hear echoes of Hebrew prophets like Isaiah and Amos in this polemic. It is a jeremiad against false idols. The state is portrayed as a monster or machine of death. Like many of the talks of the book's first part, it is structured around the life/death binary. The state brings death; life begins where the state ceases. Just as life/death is an absolute duality, so one is either inside or outside the state. The boundary line is not in question. In this respect, Zarathustra apparently agrees with the state's self-definition, which insists on its authority within clearly delineated territory. Is the state's death-making too obvious to mention? Many of Nietzsche's close readers do not even remark on the emphatic and repeated claim that the state is a death-dealing monster.[22]

What makes the state an agent of death? Zarathustra does not speak explicitly here of the state's power to kill in war, extending both to enemies and to subjects, who may be ordered to risk or sacrifice their lives. The closest he comes to such a discussion is "On the Pale Criminal," addressing the "scarlet judges," who cruelly enjoy sending the murderer to be executed (*Z* I.6). Here the murderer not only finally admits legal guilt but, more importantly, recognizes that his conscious intention (that of the lesser self or ego) was a self-deception, for the desire of the greater self was to kill. So Zarathustra's listeners or disciples have already been told that the officials of the state are using the appearance of honor it gives them (in their scarlet robes) as a conscious or more likely unconscious device for their own bloodthirsty desires.[23] "On the New Idol" is directed to those in the state. You are either in it or outside it: "In some places there are still peoples and herds left, but not with us, my brothers: here there are states." So, in the largest relevant context, call it the earth, there are at least three forms of human organization: peoples, herds, and state humans or *Staatmenschen*, those who live within the world of the state. I use the word "world" advisedly, both to preserve a certain relation between world and state, and to allow that within the world of states there are expats, immigrants, and exiles (all defined in relation to the state), as well as state officials and even hereditary monarchs and tyrants.

Unless we see the anthropological or geophilosophical pluralism presupposed here, we will not understand the force of Zarathustra's definition of the state as a wielder of death, including not only the most obvious forms just cited, but the threat of death, and the encouragement of living death or "slow suicide." "The state? What is that?" So Zarathustra puts the question. And here comes the definition:

> Well then, now open your ears for me, for now I say to you my word about the death of peoples.
>
> State is the name for the coldest of all cold monsters (*Ungeheuer*). Coldly it tells lies; and this lie crawls (*kriecht*) out of its mouth: "I, the state, am the people."

Hearing the state described as a "cold monster," it is difficult not to think of the Hobbesian Leviathan. Seeing what Zarathustra means by the death of peoples requires uncovering the Leviathan's lie as a lie. The state kills its own people through sending them to war, committing genocide, or executing them, and through soul-murder as it deceives a people into confusing itself with the state. If a people swallows this lie it is lost. Once a *people*, it now becomes the *population* of a state claiming authority over their lives. The state as we moderns know it practices biopolitics by transforming people into population. So far as they are members of the population, no longer rooted in the life of a people, they become "many-too-many (*Viel-zu-Vielen*)" or "superfluous people (*Überflussige*)." The point here is not to distinguish superior and inferior classes or ranks within the state but to suggest that all those who accept their position in the state are, as such, simply elements of population. Zarathustra had denounced the last humans for subjecting life to measurement, which is fundamental in the state's relation to population. Life subjected to measurement, calculation, and control is, Zarathustra claims, a kind of death, a living death ("slow suicide"), if you will. Biopolitics is necessarily thanatopolitics. Michel Foucault observed that traditional political power could be characterized as letting live and making die, while modern power is defined rather by making live and letting die.[24] Whether managing the population's public health and education by means of discipline and surveillance or sifting billions of electronic messages and transactions for market research or detection of political deviants, population is analyzed, shaped, and actions channeled through quantitative reduction. This analysis, shaping, and channeling is the modern way of making live. It is a shaping and measuring that devitalizes.[25]

"Far too many are born; for the superfluous was the state invented!" All are superfluous. The greater the numbers, the more zeroes both numerically and metaphorically. As superfluous, they are on the way to the condition of "bare life" as conceptualized by Giorgio Agamben. The state is happy to have surplus population. It is a monstrous apparatus for manipulating the surplus through staging a competition for riches and power. That great cynic Hartmann had said pretty much the same thing, although he thought this was simply the inevitable final stage of the *Weltprozess*. Indeed, Hartmann drew the same conclusion as Nietzsche: a life managed by the modern state, in which all are subject to its economy, are virtually dead. They must eventually see that this last chance for human happiness is hopeless, so as good Schopenhaurians they will all die by choice, exhaustion, or mutual destruction. Peter Sloterdijk elaborates the same theme when he argues that the early modern state encouraged overpopulation. Seventeenth-century witch hunts, coinciding with the rise of the absolute state, were directed against midwives, who frequently shared contraceptive information and helped to terminate unwanted pregnancies. This suppression led to excess population, which could then serve statist economic and political interests.[26]

With a bit of nostalgia, Zarathustra observes that once upon a time, and perhaps still "elsewhere," there are or were peoples who received love and faith from the creators of the people's values. Without such originary creative force there is no people. The annihilators who set up states "hang a sword and a hundred desires over them all." The sword is the ultimate threat of death, and the desires are those promoted by the state to bind the population to the way of life, or "slow suicide," that feeds the monstrous apparatus (think of Hobbes's absolute sovereign and his idea of the "commonwealth"). The state engages in systematic linguistic confusion, exchanging and perverting the notions of good and evil, presumably the good and evil established by the creators. Such confusion is the sign (*Zeichen*) of the state. "Verily, the will to death is what this sign signifies. Verily, it beckons to the preachers of death." So, behind the primal lie by which the state tries to pass itself off as the people and its confusing transvaluation of the people's values, we can detect a "will to death," a will that allies the state with "the preachers of death." Zarathustra had already devoted a speech to these preachers. The state reveals its "will to death" by allying itself with them. The state is imbricated from the beginning with such preachers; we might say that it relies necessarily on an explicit or implicit political theology. Whether they preach death directly (like Hartmann and the Buddhists) or indirectly, say, by preaching the superiority of "eternal life" to earthly, these preachers support the state's ideology.

We might be surprised to find Nietzsche apparently defending the integrity of traditional values of good and evil against their confusion by the state. This must be understood within the context of his denunciation of the state for destroying peoples. In "*On the Thousand and One Goals*," Zarathustra maintains that such tables of good and evil are essential to the life of a people. Yet the speech on the state does not encourage a return to the traditional life of a people, a people already devitalized and rendered superfluous population. No, Zarathustra enjoins his listeners to "shatter the windows and leap into the open air!" because "free for great souls the earth still stands even now." Don't walk through the doors, we might say, because there we have to pay tolls and have our documents stamped by the Leviathan's functionaries. Freedom, as Foucault says in another context, lies in a more radical escape that breaks through the windows.[27] What lies beyond the walls of the state is not the world, a term too closely associated with Hegel's claim that the world is the totality of states, but the earth, the self-renewing site of mobile, experimental human habitation. On the earth are rainbows and bridges that may lead to the *Übermensch*. In *Zarathustra*'s rhetorical and poetic discourse, we should not expect specific directives as to whether escaping from the state requires leaving its territory or is compatible with internal exile, modest poverty, and solitary or coupled life (Zarathustra speaks of the places that await *Einsame und Zweisame*).

After his series of town speeches, Zarathustra leaves his disciples and returns to his cave. From this point on, landscape, sea, and sky become more insistent dimensions of the story, both as setting and as subject of thematic exploration. We learn in Part II that Zarathustra's "wild wisdom became pregnant on lonely mountains; on rough stones she gave birth to her young . . . Now she runs foolishly through the harsh desert and seeks gentle turf" (*Z* II.1). This "wild" wisdom, outside the striated space of the state, inspires a journey of sea voyages and island hopping. These range from sightings of the isle of the dead to a volcanic atoll, and include sojourns on the "isles of the blessed"—adapting a classical motif, which might lead us to think of "bubbles" for the affluent and sophisticated. Geographically, the setting is an archipelago, a series of islands with distinctive attractions and problems. They are the singularities of the earth.

Seas, deserts, caves, and mountainous terrain all become forms of Deleuzian smooth space, fields of unconstrained movement of bodies and thought. In "On the Land of Culture (*Bildung*)," Zarathustra presents himself as a wanderer against his will, driven out of cities, fatherlands, and motherlands. He thus rejects the nation-state, a place that would claim its inhabitants by a parental model of natality, for the sake of a new people and new earth, called his "*children's land*, yet undiscovered in the most distant sea" (*KSA* 4.155).

"On Great Events" deploys the figure of a volcanic island in addressing the politics of the earth.[28] The title alludes to Hegel, who defined great events as those decisive for the world-historical development or transformation of the state; it is also a rewrite of Nietzsche's first stab at formulating his own version of the great event in *UO* IV. Commentators typically read the chapter as directed against state-oriented propagandists and politicians, especially Rousseauian enthusiasts of revolution.[29] These are allegorized as the "fire-hound," who is one of earth's "skin diseases." The "figure" of Zarathustra was seen to arrive at this Stromboli-like island by flying over the sea. The volcano (literally a "fire-mountain") that his simulacrum entered is said to be the gate of hell in local legend. Once he reassures his disciples that he is safe (only his shadow descended into volcanic hell), Zarathustra relates his dialogue with the fire-hound, an ego puffed up with a desire to expand its power, a rebel or revolutionary.[30] Such fiery demagogues are said to be at most "ventriloquists of the earth," producing the illusion of a politics that speaks from the ground of being. They give the impression that the earth, as reterritorialized by the state, constitutes a nation's true identity. In effect, they buy into Hegel's conception of world-history as the story of states. The secret unknown by the fire-hound (and the state-philosophy he represents) is that "*the heart of the earth is of gold.*" This explicitly geographical and geological chapter insists that the resources of the *Menschen-Erde* are rich in possibility. It is constituted by passionate human bodies, their combinations, and their transformations on, by, and through the earth.

Zarathustra speaks to his disciples, who anxiously worried whether it was indeed he who plunged into the "fire-mountain." While they voice doubts about the old wives' tale that it is the gate of hell, "in the ground of their souls they were all of them filled with concern and yearning." Zarathustra takes their disturbance as a teachable moment. Those who saw the figure's descent or heard the news may think they have witnessed a spectacular event, a "great event." Zarathustra's lesson to the disciples is that such spectacles are not great events, just as he explains to the fire-hound that neither are they constituted by rebellions and revolutions. His implicit message is something like this: "You thought that was a great event? Let me explain what a great event is and is not."

In hearing Zarathustra correct both his disciples and the fire-dog on their superficial idea of great events, we should recall that Nietzsche once thought that he saw a great event of the earth in Wagner's supposed severing of East and West. Zarathustra's discourse to his disciples—those who have presumably pledged their loyalty to the earth—is a story of elemental love and strife, a geology of morals.[31] As he explains: "The earth . . . has a skin, and this skin has diseases. One of these diseases, for example, is called '*Mensch*.' And another one of these diseases is called "fire-hound." The earth is a complex of strata; we tend to ignore its intricate stratigraphy because we are deafened by the voice of the fire-hound and his kind, who see only surfaces. The earth's interior is mineral, sometimes molten. Humans interfere with the biosphere (the skin) in many ways, mineralizing the surface in stone, concrete, and brick, and releasing noxious materials into its atmosphere. Today they pave the earth with highways and parking lots for oil- driven vehicles. They turn the earth inside out seeking fossil fuels and minerals. This surface disorder, which bites into its ground, is like an autoimmune disease, the earth being disrupted by its own products as they war for resources. To understand the rebellious fire-hound, Zarathustra conducts a full analysis of its elemental dimensions: not only fire, but sea, smoke, and earth (mud). Now he knows that the fire-hound's noise is not a "great event." Politics as generally practiced is all smoke and mirrors. Political rebels accept an unquestioned context of state and self, limiting their contestation of received values. Zarathustra recounts his conversation with the fire-hound: "'Freedom' is what you all most like to bellow; but I have unlearned my belief in 'great events' whenever they are surrounded by so much fire and smoke." The struggle for freedom, of course, is what Hegel sees as the prime engine of world-history, that which brings about "great events." But genuinely great events arise in those "stillest hours" when we become aware of new values. In contrast, the state (or churches, those quasi-states) are fire-hounds, bellowing ventriloquists of the earth proclaiming their own absolute importance. Another fire-hound, Zarathustra continues, "really speaks out of the heart of the earth. He exhales gold and golden rain." This beast is at home in elemental extremes, like "ashes and smoke and hot slime," but Zarathustra knows that "*the heart of the earth is of gold*." Gold, Zarathustra's disciples know, is glowing and giving, the sign of "the gift-giving virtue (*schenkende Tugend*)" (*Z* I.22). The molten, radiant core of the earth has an affinity to the sun. Geological flow is what the earth is about. Ordinary politics takes place among skin diseases, the superficial flows of humans and states. On the horizon is the project of a geology of morals, a project sketched by Deleuze and Guattari. "Who does the earth think it is?" is a question that must be taken seriously, whenever there is a discourse of nation, national identity, and where ownership of land and the limits of ownership are at issue. Zarathustra's claim that the heart of the earth is of gold is rephrased in *A Thousand Plateaus*, by Professor Challenger, who

> explained that the Earth—the Deterritorialized, the Glacial, the giant Molecule—is a body without organs. This body without organs is permeated by unformed, unstable matters, by flows in all directions, by free intensities or nomadic singularities, by mad or transitory particles.[32]

"On Great Events" once more problematizes the Hegelian claim that great or world-historical events are exclusively those having to do with the development, transformation, or conflict of states. Humans, now a skin disease on earth's surface, could transform their habitat into a truly human-earth, more specifically a garden (as we hear later). What would it mean to be loyal to the earth, to raise the question of its meaning and direction with the passion so far channeled into state and church? Empedocles gave an enigmatic answer when, as tradition has it, he jumped into Etna; at the end of his talk to his disciples, Zarathustra insists it was only his shadow or specter that mimicked that act. Yet he is puzzled by the specter's reported exclamation: "It is time! It is high time!" Time for what? For a great event involving the earth? The question hangs in the air although the next chapter introduces more doubt with the figure of "The Soothsayer." That symbol of Schopehauerian pessimism challenges any hope raised by Zarathustra's loyalty to the earth and its golden heart, teaching that earth is now exhausted—fruit turned rotten, soil cracked, wells dried, wine poisonous, sea become shallow swamp. That picture of the earth then motivates Zarathustra's nightmare of himself as a night watchman of universal death, a dream that makes a volte-face into a revelatory vision of revived earthly life when the glass coffins containing the dead burst open with raucous, carnivalesque life.

Part III is the story of a homecoming by walking, sailing, and mountain-climbing. As Zarathustra ascends a mountain ridge, a new perspective opens on a fresh sea, and he reflects on the alternation of perspectives, which turns things upside down, and the need to discern grounds and backgrounds, not merely the foregrounds that appear to overly obtrusive eyes. He disorients the framework of the picturesque poem or landscape painting with its single fixed angle on the earth. Zarathustra invokes geological time and space to suggest a general pattern of analysis:

> Whence come the highest mountains? I once asked. Then I learned that they came out of the sea. The evidence is written (*geschrieben*) in their rocks and

> in the walls of their peaks. It is out of the deepest depth that the highest must come to its height (*KSA* 4.195).

These strata are forms of writing, codings of the earth; this provides a model for understanding how human singularities emerge from their own depths and from reading the inscriptions of their movements and speeds. The earth is a text that we must learn to read.

How does Zarathustra read the earth? Emerging from his struggle with the "abysmal thought" of the last human's eternal recurrence, he confesses that the human-earth had seemed to turn into a cave of death and decay. Convalescing from this agon, he accepts his animals' cheering news that the world awaits him as a garden (*Z* III.13). The garden is a richly textured Nietzschean concept, already prominent in *Dawn* and *Gay Science*. Later we will ask more pointedly whether the garden earth fulfills Nietzsche's aesthetic-political vision of earth's transformation. Concluding the third part of the book, which Nietzsche seems to have intended once as the conclusion of the whole, Zarathustra sings his celebratory song of the earth, "The Seven Seals" (*Z* III.16). This geophilosophical rhapsody should be read in opposition to the apocalyptic geography of the otherworldly Biblical *Apocalypse*, with its conjuring of earthquakes and the end of the world. In contrast, this anti-Christian paean to the earth celebrates the earth's immanent divinity, as in the third stanza:

> If I ever played dice with gods at the gods' table, the earth, till the earth quaked and burst and snorted up floods of fire—for the earth is a table for gods and trembles with creative new words and gods' throws: Oh, how should I not lust after eternity and after the nuptial ring of rings, the ring of recurrence?

Each stanza of "The Seven Seals" is an answer to the question of what the meaning or direction of the earth shall be; the whole is a vision of smooth space that must be heard as a response to the hypermeasured space and time of rapid globalization that makes the earth small, breeding last humans who hop like fleas: "The coast has vanished, now the last chain has fallen from me; the boundless roars around me, far out glisten space and time; be of good cheer, old heart!"[33]

Part IV of Zarathustra's story finds him playing host in his mountain aerie to eight "higher men," who are driven by distress to seek him out. These aspirants to a meaningful life on the earth can be read as allegorical representations of artists, ascetics, scientists, rulers, and others in quest of earthly wisdom. The aged guru transforms his solitary redoubt into a landscape of celebration as he oversees a festival of the earth, on solar time. We could think of this end of Nietzsche's philosophical landscape poem (which may, chronologically speaking, be an interlude) as Zarathustra's garden party, on the transformed earth that his animals had described as awaiting him as a garden.[34]

BEYOND PEOPLES AND FATHERLANDS: NOMADS, HYBRIDS, AND THE CENTURY OF THE MULTITUDE

If *Zarathustra* is a phantasmatic and hallucinatory landscape poem, *Beyond Good and Evil: Prelude to a Philosophy of the Future* surveys the thought and culture of late nineteenth-century Europe with a view to how philosophers of the future—future philosophers, but more importantly thinkers of futurity and the event—can carry forward the European adventure. Zarathustra's parables are replaced by aphorisms on the cultural heritage and emergent prospects of "the free spirit" and "the good European" (a figure that should be read in opposition to both national identifications and the "last human"). The two books constitute parallel responses and experimental intensifications of questions such as: How does thought deterritorialize and reterritorialize itself? What does it means to think on and with the earth? What has the age-old absolutist dichotomy of good and evil meant for life on the earth? What might the earth be like if the moralities and religions empowered by that opposition ceased to be dominant?

Some critics read Nietzsche's earth as parasitic on his longing for the Greek world, mediated by Hölderlin, his early poetic model. Others take a phenomenological approach, either celebrating his aerial fantasies of free-flying birds (Bachelard) or questioning his supposed avoidance of the all-engulfing sea (Irigaray). A different perspective emerges when we contextualize Nietzsche in terms of his concern with the study of human geography as it was developing in the nineteenth century. For example, he read and annotated Friedrich Ratzel's *Anthropo-Geographie* (1882), in which the latter argues that philosophers and geographers, notably Kant and Hegel, have subordinated geography to history; yet in order to take the broadest view of human life on the earth, we require a global perspective recognizing that written history provides only a small sliver of relevant data about the earth as humanly inhabited. Ratzel emphasizes the limits of a perspective that takes the recorded history of states as the center of our understanding of human habitation and movement, marginalizing migrations, wanderings, and climate and environmental changes (cf. *BGE* 32, *GM* II). Recalling this context provides an alternative to an overly interiorized, subjectivist, and "existential" reading of Nietzsche. Moreover, it allows us to focus on Nietzsche's critical geophilosophical analysis of Germany and of Europe.

Europe is a constant theme, perhaps the major theme of *Beyond Good and Evil*. Europe's future is what is at stake. Nietzsche is on the lookout for "good Europeans," and he takes on the task of providing a critical analysis of Europe's desire to become one. Yet what is Europe? As Nietzsche explains several times, he has in mind much more than the conventional geographical division that marks off this "little cape of Asia" (*WS* 215, *BGE* 52, cf. *BGE* 208); it is necessary to understand Europe in the widest context, with an "Asian and trans-Asian eye" (*BGE* 56). Europe can be understood genealogically as including all the regions whose cultures owe a fundamental debt to Europe in this narrower sense; so for him it includes North and South America. Unlike Hegel, Nietzsche does not define Europe in terms of its supposed destiny to establish a certain kind of political state. Europe is in crisis—whether it knows it or not—as it struggles with the collapse of Christianity, the emergence of democratic attitudes and practices, the threat of nihilism, and the possible rule of the herd and the last man. As Zarathustra challenged his auditors to take responsibility for the direction of the earth, the Nietzsche of *Beyond Good and Evil* and subsequent works interrogates Europe in terms of its largely unconscious response to these challenges, while advancing the possibility that "Europe" may flourish anew in unanticipated ways.

Nietzsche's engagement with Europe's future can be understood from the perspective of what Deleuze and Guattari (hereafter Deleuze) call geophilosophy, a field of thought they credit him with inventing.[35] Geophilosophy, says Deleuze, recognizes that thinking goes on not between subject and object, but "takes place in the relationship of territory and the earth."[36] Thought, whether philosophical or prephilosophical, involves a process of territorialization, deterritorialization, and reterritorialization through which it takes up a variety of positions in relationship to the plane of immanence, understood in its most immediate and familiar manifestation as earth, the ground of life and thought. To give a necessarily skeletal summary: We humans (and all living things) territorialize by staking out a space, a place: we settle down, we cultivate a field, we mark the borders of our situation, whether in the areas traced by the Australian aborigines in song lines, or the homeless person's little stretch of sidewalk or space under the bridge. Deterritorialization consists in an idealizing movement by which actual, physical space becomes subsumed within some structure requiring a more conceptual definition. A political state, an empire, declares that the meaning of a certain assemblage of people, land, and resources consists in a unified structure. Ancient Athens, as it showed in the battle of Salamis, was able to conceive of itself in deterritorialized fashion by configuring itself as a mobile political structure, not absolutely tied to a fixed place. Think of reterritorialization as a "back to the land" movement, the reclaiming of a territory that had previously been absorbed by a deterritorialized entity. Although Deleuze and Guattari do not limit the use of these notions to their most literal applications to earth and the land, this is surely one of their primary senses, and I will follow it here. Deleuze proposes that the philosophical project is one of absolute deterritorialization (which we could read as the way in which they take Plato's ambition to contemplate all time and existence [*Republic* 486a] as applicable to a world of immanence). Yet philosophy reterritorializes itself, three times, they say. First on the Greek maritime world of commerce and city-state, whose figure is the friend. Second, on modern capitalism and the institution of the national state. And, proleptically, on the "new people and new earth" that Deleuze claims is emerging and that current thought is, sometimes unconsciously, attempting to evoke.

Beyond Good and Evil invites the reader to identify herself as a "good European" who will give a new sense to life in Europe, which has long since deterritorialized itself from the "little cape of Asia." Indeed, we won't understand Nietzsche's project at all if we begin with a human geography that fails to see Europe in motion. Europe then is not to be comprehended as simply a plurality of states with stable constitutions grounded in religious and ethnic identities (Hegel's version of the Westphalian model) but in terms of mobility, difference, and multiplicity. It is necessary to think beyond "peoples and fatherlands" (the title of the book's most ostensibly political chapter), and to evoke this sense of being on the move Nietzsche once again returns (as in announcing the death of God) to an astronomical metaphor: "I hear with pleasure that our sun is swiftly moving toward the constellation of *Hercules*—and I hope that man on this earth will in this respect follow the sun's example? And we first of all, we good Europeans!—" (*BGE* 243).

The diagnoses of European philosophy and art in *Beyond Good and Evil* offer a critical account of how thinking takes place in the relation between earth and territory. It is a geo-logic, a cartography of human constructions of the *Menschen-Erde* and an evocation of their futurity. The problem, we will see, that underlies Nietzsche's interrogation of "Peoples and Fatherlands" is that philosophy, despite its universalistic ambitions and pretensions, is unconsciously territorialized. It *aims*, as Deleuze says, at absolute deterritorialization, but necessarily reterritorializes itself, and in modernity this means that it produces itself in forms associated with the national state.

The "Preface" to *BGE* recalls that monstrous forms of thought like "astrology and its 'supra-terrestrial' claims" have had the most stupendous effects on the lived earth, as in "the grand style in architecture in Asia and Egypt"; Nietzsche ranks these earliest monumental architectural forms among humanity's fruitful errors. Pre-Copernican—they see the stars as divine, taking a local capital to be the center of the cosmos—Nietzsche nevertheless admires these prime inscriptions on the earth as grand experiments in giving a direction to the earth. They are the architectural signature of thought still tied to transcendence, and as such are necessarily figurative and diagrammatic. These are ancient attempts to comprehend the meaning and direction of the earth. Both dogmatic philosophy and "the grand style of architecture" demonstrate that "all great things first have to bestride the earth in monstrous and frightening masks in order to inscribe themselves in the hearts of humanity with eternal demands" (*BGE* "Preface").

Here it is helpful to recall Deleuze's discussion of figurative or diagrammatic philosophy in his chapter on geophilosophy. He asks, "Can we speak of Chinese, Hindu, Jewish, or Islamic 'philosophy'?"[37] The answer is conditional: "Yes, to the extent that thinking takes place on a plane of immanence that can be populated by figures as much as by concepts." Ultimately, however, there is a distinction between figure and concept, however difficult it may be to discern in specific cases. "Figures are projections on the plane, which implies something vertical or transcendent."[38] Although Deleuze does not mention Derrida in this connection, I read this distinction of figure and concept as an implicit critique of the politico-aesthetic argument in *Of Grammatology*, which aims at undermining logocentrism by demonstrating the omnipresence of writing, and so undercutting the ethnocentrism that distinguishes peoples with and without writing, or alphabetical scripts from inscription in general. Gayatri Spivak has criticized Deleuze for ethnocentrism on just this score.[39] Whatever we might think about this claim, however, note that Deleuze distances himself from Hegel and Heidegger, who find the beginnings of philosophy in the original nature of the Greeks. Rather, philosophy arises as an accident of geography: "The birth of philosophy required an *encounter* between the Greek milieu and the plane of immanence of thought." Thought proceeds even under figurative forms, but without the relative deterritorialization of the maritime culture of distinct cities as opposed to empire, there would have been no friend, no *philos*. Without the friend, there is no notion of philosophy as common activity, but only of the radically marked individual philosopher as an idiosyncratic phenomenon. It is this that Nietzsche has in mind in *Philosophy in the Tragic Age of the Greeks*, when he says that without a common vibrant culture philosophers can appear only as isolated comets streaking through the sky. It is the Greek maritime world that provides a milieu for the friendship that is necessary for *philosophia* as contrasted to the isolated *philosophos*.

"Peoples and Fatherlands," even more intensely than the rest of *Beyond Good and Evil*, offers an inventory of forms of territorialization and its variants. This chapter interrogates its title concepts, as well as the state, empire, and addiction to the soil: the national characteristics of German, French, and English thought. It contains an analysis of Europe and the Jews, and points to the emergence of supranational and nomad peoples. Nietzsche asks whether and how "the pathological estrangement which the insanity of nationality has induced" could be relieved by "Europe's desire to become one" (*BGE* 256). I will read the chapter with an eye to understanding its implications for what Nietzsche calls the direction of the earth. In "Peoples and Fatherlands," the concluding aphorism announces that "*Europe wants to become one.*" It is an italicized announcement. What is it to become one? Is not Nietzsche in many ways the enemy of "the one"? I think we are right to be suspicious of any possibility of essentialism here, and we might wonder if Nietzsche has forgotten the lessons about the *Weltprozess* he had given to Hartmann. We can best approach the question of the one, the European one (and the multiplicity that is its apparent other), by way of the geophilosophical concepts that Nietzsche deploys here.

This chapter should be read as a coherent argument, not simply as a collection of aphoristic musings and observations on European nationalities. As often with Nietzsche, the argument is somewhat indirect, even labyrinthine; nevertheless it is susceptible of analysis. What follows is an attempt to unearth the outlines of that argument, which attends to his use of names (e.g., Wagner, Goethe, Napoleon) to designate affective and cultural events. The argument proceeds to articulate the relationship he discerns between music and territory. "Peoples and Fatherlands" begins, not altogether surprisingly, with the Germans. Can't Nietzche finally get away from them, we might ask, even here where he assumes the conceptual persona of the good European? The Germans provide a paradigm case for the question of national essentialism and its alternatives, because of Nietzsche's own descent and affiliations (*Herkunft*), his earlier cultural nationalism, and the rise of German imperialism. And, even more parochially, it is his old idol—and now antagonist—Wagner who is emblematic of the Germans (*BGE* 240). Music, throughout the chapter, is taken to be a primary mark of peoples and fatherlands, as well as of new forms of thought and social formations, which are not mentioned in Nietzsche's title. Wagner's music had become a national symbol, a development Nietzsche had attempted to accelerate from 1869 to 1876, until his crisis at the Bayreuth festival. Both Germanophiles and Germanophobes, then and since, have taken Wagner as a signature expression of German nationalism. Nietzsche's diagnosis of this putative hyper-German music, he says, is inspired by "hearing once again for the first time" the overture to the *Meistersinger*. This "once again for the first time" forces us to ask *what* is newly heard (and why is the music always heard as if for the first time)? First, the music's *history*: Wagner "is magnificent, overcharged, heavy, late art that has the pride of presupposing two centuries of music as still living, if it is to be understood." Wagner offers an amazing range of internal variations, from "pompous-traditional" to "saucy," from "archaic" to "young," shifting from a "broad and full flow" to moments of "inexplicable hesitation," and all of this is colored by a "current of well-being" and "the artist's happiness with himself."

So far so good, it seems. Wagner offers, then, a form of monumental history. But all such monumentality is also an instance of that "semiotic compression" that, as Nietzsche says, must be analyzed genealogically rather than through essentialist definition. Launching that analysis, he notes what is *absent* in this "comprehensive" music: "no south . . . no dance, scarcely any will to logic." The ambition to produce a musical summa leads to "something German in the best and worst senses of the word, something manifold, formless, and inexhaustible in a German way." If music is the hermeneutic clue to what it means to be German, we find no clear core of sensibility and creativity, but indeterminacy and vagueness. Where we thought we would discover a firm national character, we find ourselves grasping at clouds. Is there no firm ground, no foundation, no essence of the German people? (Recall that Nietzsche previously dismissed the state as a fragile construction held together by a state of emergency based on manipulating peoples' fears—[*HAH* 472–75; cf. *BGE* 201]). Are peoples and fatherlands as unstable as the state?

In section 244, Nietzsche expands his initial weather report on the Germans by means of an antiessentialist analysis. He begins by saying that they have often been taken to be profound or deep (*tief*), to have, we might say, a deeply grounded foundation. The point has often been made about the basic metaphors of German philosophy. The importance of the *Grund* is obvious in Kant's architectonic, in Heidegger's search for the way back to the ground of metaphysics. Nietzsche claims that the appearance of depth is grossly misleading, and can be cured by a "little vivisection" of the German soul: "More than anything the German soul is multiple (*vielfach*), of diverse origins, more piled up and pieced together than actually constructed: that is due to its heritage (*Herkunft*)" (*BGE* 244). The German soul is a multiplicity, far beyond even the duality acknowledged by Goethe's Faust, who spoke of two souls dwelling in his breast. This is not surprising when we recognize its variant strands and observe its assemblage from diverse components. Foucault reads Nietzsche, especially in the *Genealogy*, as typically distinguishing origin (*Ursprung*) and heritage (*Herkunft*). While his reading of the *Genealogy* text may be somewhat tendentious, the pluralizing impulse he discerns is active here in the account of the complex German soul, and frequently in other contexts[40] Here Nietzsche pluralizes the "origin," thus in a certain sense undermining the very notion of origin. A genealogical analysis of *Herkunft* will show us what an unstable assemblage it is. Genealogy reveals more differences, for the Germans are "a people of the most monstrous mixture and medley of races." They are "people of the middle" not simply with regard to their geographical place in Europe, but because they are a crossroads, an area of intersection, of so many currents: "The Germans are more incomprehensible, comprehensive, contradictory, unknown, incalculable, surprising, even frightening than other peoples are to themselves: they escape *definition* and would be on that account the despair of the French" (*BGE* 244). The tool for dispelling the national "shadow of God" (*GS* 108) is genealogical analysis, showing, as in a lapidary formulation of the *Genealogy of Morality*, that only that which has no history can be defined (*GM* II.12).

Yet if the Germans are not profound, how do we account for their reputation? The effect of depth is created by a temporalization of space, a translation of geography into history. We get a new account in architectural and (once again) meteorological terms:

> The German soul has its passageways and interpassageways; there are caves, hideouts and dungeons in it; its disorder has a good deal of the attraction of the mysterious; the German is an expert on secret paths to chaos. And just as everything loves its metaphor, the German loves clouds, and everything that is unclear, becoming, twilit, damp, and overcast; whatever is in any way uncertain, unformed, blurred, growing he feels to be "profound." The German himself *is* not, he *becomes*, he "develops" (*BGE* 244).

The German translates space—whether understood architecturally, geographically, or meteorologically—into time. Nietzsche says that this is what Hegel brought into a philosophical system and Wagner set to music. For Hegel, geography is *aufgehoben* in world-history, as the world-spirit inexorably travels from east to west, a movement that itself transforms the daily and cyclical apparent movement of the sun into a historical development. While the east/west binary is convenient for translating geography into history, Nietzsche typically challenges it by invoking the north/south axis.

Germany is a crucial site because, on the one hand, its indeterminacy (its cloudiness, its commitment to becoming rather than being, its ability to tolerate and even celebrate contradictions) could render it hospitable to the new thinking demanded by the European democratic movement. On the other hand, the productive possibilities of this multiplicity are at risk of being foreclosed by the *systematic* form imposed on it by Hegel and his like (including Wagner). Consider the form of this systematicity. Hegel's philosophy of history is the antithesis of the geophilosophy that Nietzsche is developing. Hegel forces becoming into a story, a narrative with beginning, middle, and end. His philosophy of history involves the *Aufhebung* of geography. In Hegel's story, the movement of history and spirit is from east to west, a single irreversible "development." In "Peoples and Fatherlands," Nietzsche counters this story with a geography alert to the constant tension and interplay of north and south. The north/south leitmotif sounded repeatedly in this chapter is to be understood as a fundamental "re*orient*ation" of how we construe the relation of thought and the earth. The French north/south fusion is therefore important as another experiment in creating human types. There is a French "taste of the spirit" that overcomes cultural nationalism (*BGE* 254). Deleuze also thinks of the south, metaphorically, as the direction that discloses the instability of conventional binary east/west political conceptualizations.[41] The thought of the "philosophers of the future"—the thinkers of futurity—does not imitate the course of the sun in order to complete a story that liberates it finally from its ground; this thought explores the north/south becomings of a Europe that is moving otherwise. Nietzsche rejoices in the ability of the French "to fortify themselves against the awful northern gray on gray" (*BGE* 254): this seems to be another allusion to Hegel, this time to the owl of Minerva, which paints its last grisaille picture in the German fog.[42] The opening to the south displaces the east/west historical axis and gives the French promise for nurturing philosophers of the future. The French can still accommodate "those rarer and rarely satisfied people who are too far-ranging to find satisfaction in any fatherlandishness, and know how to love the south in the north and the north in the south—the born *Mittelländer*, the 'good Europeans.'"

EXCURSUS ON MUSIC

It is here that Nietzsche praises Bizet for having "discovered a piece of the *southernness of music*" (*BGE* 254) and proceeds in the next aphorism to imagine a "supra-German music," a music "redeemed from the north." It is understood geographically:

> [It] does not fade, yellow, or pale at the sight of the voluptuous blue sea or the luminous Mediterranean sky . . . a supra-European music that still stands its ground before the brown sunsets of the desert, whose soul is related to the palm tree and that knows how to wander and to be at home among huge, beautiful, lonely beasts of prey (*BGE* 255).

Are these remarks simply expressions of a wish for a different kind of program music, to replace northern pastoral symphonies with southern desert-and-sea compositions? While much of "Peoples and Fatherlands" deals with what seems to be the most ethereal, least earthly of the arts, music, this is not inconsistent with its geo-logical orientation. Here the body's musical expression is also a song of the earth. Consider the contrast of Nietzsche's concept of music with Hegel's. For the latter, music is the art that definitively breaks with space and externality; it unfolds the inward realm of the spirit in time.[43] Nietzsche turns the tables on Hegel, arguing that his philosophy and Wagner's music (which he calls its artistic equivalent) must be understood geographically; speaking of this pair, he explains their similarities by claiming that "the German loves clouds and everything that is unclear, becoming, twilit, damp, and overcast" (*BGE* 244). The claim to transcend and sublate geography in Hegel's philosophy of history is itself a function of a cloudy climate of ideas.

Deleuze's concept of the refrain as a marking of territory makes a structurally similar if more explicit attempt to understand how music establishes a meaningful space.[44] This account, like that in "Peoples and Fatherlands," is a geophilosophical analysis emphasizing the interrelations of music, territory, and the political. Deleuze distinguishes three typical forms (not successive stages) of the relation between sonority and territory. (1) A child might sing to himself in the dark, creating an elementary if fragile sense of safety and shelter; this is what they call the *classical* mode, aimed at wresting order from chaos. (2) A wall of sound may be created in order to provide the sense of a more permanent home, constructing an interior space; Deleuze calls this *romantic*, describing it as the search for a genuine territory. (3) Music can begin to open up toward the outside, to the future and the cosmos. This *modern* form is adventurous and exploratory, insofar as it is not irrevocably tied to a specific territory. Deleuze cites Nietzsche's refrain of eternal return as a model of cosmic music, thus recognizing its contribution to the *Sinn der Erde*.

MOTHERLANDS, FATHERLANDS, MULTITUDES

This analysis of music throws some light on the figure of the "good European" that Nietzsche develops here by reading national and supra-national modes of thought and feeling through musical styles. The good European is not (certainly is not *primarily*) the citizen of a single new European state. The good European is the goal of *Beyond Good and Evil*, holding the same place in that work that the posthuman occupies in *Zarathustra*. If Zarathustra is a fantastic figure, deriving from a specifically non-European landscape (Persia), the good European is something of tomorrow or the day after. We do not have to wait for the "great noon" for the emergence of the good European; this is the name of the future that is arriving.

In the second aphorism of "Peoples," Nietzsche speaks of "dull and sluggish races who would require half a century even in our rapidly moving Europe to overcome . . . atavistic attacks of fatherlandishness and gluing themselves to the soil (*Schollenkleberei*)" (*BGE* 241). This is a reactive reterritorialization that cannot acknowledge itself. This dogged geographical essentialism obstructs movement and stands in the way of the "dull races'" expansion or development from other "races," such as the possibility that Nietzsche (the supposed proto-Nazi) projected of Germans and Jews forming a new hybrid. (*BGE* 251). Cloudy indeterminacy is a strength, if it enables receptivity to the exterior, but in gluing the nation to the soil, sticking it in the mud, unnamed statesmen (Bismarck and his like) narrow the spirit and degrade taste. Unspoken here is the tension between "people" and "fatherland." A people is fundamentally mobile and active, although "dull and sluggish" by turns. It handicaps itself by assuming a national identity through hypostatizing its geographical situation into a "fatherland." Peoples are experiments with a future, not essences to be preserved. Here we might think for a moment of a suppressed or implicit contrast between *fatherland* and *motherland*. "Fatherland" suggests singular and patriarchal authority, the daddy state that tends to claim emergency authority, declares a state of exception, and maintains its position by a propaganda of fear. Only at one point does Nietzsche speak of motherlands, and it is simply to equate them with fatherlands. Zarathustra says that he has searched for father- and motherlands, but has failed to find a homeland: "unsettled am I in all settlements (*unstät bin ich in alle Städten*) and a departure at all gates."

He declares that he has "been driven out from all father- and motherlands" and loves only his "*children's land*, undiscovered, in the farthest sea" (*Z* II.14). Like the contemporary nomad, he is unsettled, but his wandering has a goal; Deleuze understands this as the search to reterritorialize philosophy on a "new people, new earth."

In the third aphorism, Nietzsche further contextualizes the question of nationality (*BGE* 242). What can be called "Europe's *democratic* movement" is a "*physiological*" process that is producing "a supra-national and nomadic type of human being," one distinguished by its high power of adaptation, fit to become an "industrious worker" and "multi-purpose herd animal." Nietzsche carefully notes that he says this "without praise or blame." What is the "nomadic type of human being"? The nomad should not be construed as a hermit, a solitary "existential" figure (associations that students typically make). Speaking here of contemporary nomads, Nietzsche emphasizes not their isolation and difference, but their tendency, in these circumstances, to homogeneity, to become a herd of adaptable workers (Hartmann, implicitly following Hegel, had neglected emigration and immigration in his globalization scenario). If "peoples and fatherlands" are mobile, experimental constructions, the nomads intensify this mobility, detaching themselves from states and their nationalisms. Nietzsche's trans-European eye sees mobility as primary in human habitation. He apparently sharpened this view through his reading of Friedrich Ratzel's *Anthropo-Geographie* (1882)—a book that he underlined and annotated. (Ratzel became a problematic figure in later German geopolitics, contributing to the formation of the concept of *Lebensraum*). Ratzel attempted to articulate the basic parameters of a truly global human geography, one not limited by the perspectives of states, therefore not by written history, itself linked to the state form. In the concluding chapter of his treatise, Ratzel highlights his signature thesis: the movement of peoples, their inevitable mixing and tendency to homogenization is the most constant feature of human life on the earth. He writes:

> The human being is restless, he strives for the greatest possible overall expansion, wherever natural boundaries do not sharply hem him in, and any anthropological conceptualization that does not take this restlessness into account, rests on a false foundation. Humanity must be seen as a mass (*Masse*) that finds itself in constant effervescent or fermenting motion (*gährender Bewegung*), and through this *Gährung* a great inner manifold (*Mannigfaltigkeit*) is united. This fermentation obtains in varying degrees, but is absent in no people or stage of

> culture. It has the tendency to make human beings ever more uniform, because mixing is inseparably bound up with this movement.[45]

Nietzsche adds to Ratzel's analysis that this mixing gives rise to hybrids and monsters, "exceptional types," not merely more uniform populations. This is not merely because of his high valuation of individual genius, but also follows from what he adapts from Darwin and Darwinism regarding the properties of a population. For example, Nietzsche begins *Gay Science* by saying that the faulty mathematics of the solemn, tragic "teachers of the purpose of existence" leads them to substitute one for the multiple; he looks forward to the time when we will have realized that "the species is everything, *one* is always none"; the teachers of the purpose of existence fail to understand the logic of the multiple: "for [them] there are no species, sums, or zeros" (*GS* 1). Here we should note Nietzsche's exclamation that this is "the century of the *multitude* (***Menge***)" (*BGE* 256). We should not be misled by translations that read "*Menge*" as "masses." "Masses" suggests only homogeneity; while Nietzsche does see the drive toward homogeneity in the emerging Europe, he also emphasizes the exceptional, the inventive, and the hybrid.[46] (Note that set theory, being developed at precisely this time by Georg Cantor to deal with the absolutely multiple or infinite, is *Mengenlehre*. I have no reason, however, to think that Nietzsche knew Cantor's work.)

Again in this spirit, Nietzsche challenges the natal or autochthonous dimension implicit in the national, the root of fatherlandishness.

> What gets called a "nation" in Europe today (and is really more a *res facta* [something made] than *nata* [born]—every once in a while a *res ficta et picta* [something fictitious and painted] will look exactly the same) is, in any case, something young, easily changed, and in a state of becoming, not yet a race let alone the sort of *aere perennius* [more enduring than bronze] that the Jewish type is (*BGE* 251).

It seems at first that Nietzsche depicts the Jews as an exception to the mobility and fictitiousness of the nations; but as the aphorism continues, it becomes clear that they are to be distinguished from other groups only by the *relative speed* of their movement, not by any essential characteristic. They represent a countermovement to the nomadic, since they are seeking a place to settle down and assimilate to some degree. Yet this very movement, Nietzsche opines, "perhaps already reveals a slackening of the Jewish instincts." So even this group, whom he had just described as "without a doubt the strongest, purest, most tenacious race living in Europe today," are subject to the general principle of mobility.

Nietzsche emphasizes the agonistic relation between nomads and the state (with its valorization of identity and its insane "state of exception" nationalism) even more than Ratzel. Deleuze suggests (from a reading of *Genealogy* II 16–17) that the nomads arise at the state's periphery and exist in constant tension with it. The nomads resist not only the existing empire or "despotic machine"; they resist the formation of states among themselves. Official history largely neglects the nomads or finds them incomprehensible. Deleuze offers an explanation:

> If there is no history from the viewpoint of the nomads, although everything passes through them, to the point that they are the *noumena* or the unknowable of history, it is because they cannot be separated from this task of abolition which makes the nomad empires vanish as if of their own accord, at the same time as the war-machine is either destroyed or passes into the service of the state.[47]

Earlier I agreed with Deleuze that *Schopenhauer as Educator* is an implicitly nomadic text and supplemented Deleuze's reading by recalling Nietzsche's incorporation of related themes in Emerson. Now whatever inspiration Nietzsche indirectly received from North America's mobile populations and expanding frontiers has been brought back to Europe with his observation, over a decade later, of its increasing Americanization (cf. *GS* 329).

Nietzsche is willing to make some conjectures about the way in which the European experiments (there is never just one) will go. He has at least two models, and it is not immediately obvious that they can be reconciled. The *first* is a democratizing and *homogenizing* movement leading to the adaptable worker who will allow others to organize their lives, close to Zarathustra's last man and Hartmann's globalized bourgeoisie. It makes possible the rise of tyrants, including the "most spiritual" kind. On a *second* model, Europe is the breeding ground for new forms of spiritual *hybridity*, and Nietzsche distinguishes the roles of receptivity and generation. He sees two possibly productive roles for Germany: as the cultural stimulus for France and as a marital introduction service for military families and Jews (*BGE* 251). Such hybrids, not the homogenized last men, are movements in the style of the good European. On the one hand homogenization, on the other hybridity. These two movements are simultaneous, because one is common, the other rare. Social, economic, and geographic mobility produces a multitude adapted to globalized conditions as well as exceptional new combinations brought about by a variety of causes. The good European might look like a French philosopher inhabited by German thought (say French philosophy from Sartre to Derrida) or the children of German-Jewish marriages. These hybrids are not themselves instances of a higher type, but signs of the fertility of Europe's productive ferment.

Now we can read the concluding aphorism of the chapter with its declaration: "*Europe wants to become one*" (*BGE* 256). Nietzsche accuses the "insanity" of all the nationalisms of a mendacious misinterpretation of Europe's desire. How are we to understand this desire to become *one*? If Nietzsche is a confirmed anti-essentialist (recall his critique of the national state and its *Schollenkleberei*), would a new political Europeanism, a "European Union," generate a new essentialism?

What does Europe want when it wants to become one? We should be puzzled when Nietzsche speaks of "Europe"—this diverse collection of peoples whose differences Nietzsche has been cataloging and analyzing throughout "Peoples and Fatherlands"—*wanting* anything. Yet Nietzsche reads this desire "in all the more profound and comprehensive men of this century," providing a representative list of writers, composers, philosophers, and emperors (Napoleon). Contrary to their own self-portraits, Nietzsche interprets figures like Stendhal, Beethoven, Heine, Schopenhauer, and Goethe as desiring that Europe become one. Each, he suggests, "anticipate[s] experimentally the European of the future," by practicing cultural combination or synthesis. They are all hybrids (or monsters). Each proceeds differently, and there is no grand synthesis, but different adventures. They exemplify a spirit of experimentation freed of nationalistic insanity, even if occasionally misunderstanding themselves as patriots, in moments of "weakness or in old age." Wagner is the main example. Despite his egregious German nationalism, Wagner's work, Nietzsche claims, is intimately related to the French romanticism of his youth. So he was never an *echt* German as a musician. Here is further confirmation of German multiplicity.

It is experimentation outside the self-imposed limits of peoples and fatherlands that makes these de facto cosmopolitans and hybrids exemplary. Nietzsche *seems* to say that we can learn what it means for Europe to become one by studying these figures: "It is Europe, the one Europe, whose soul surges and longs to get further and higher through their manifold and impetuous art," but he breaks off his sentence with a question "—where? into a new light? toward a new sun?" The question is left open whether these geniuses aim, even unconsciously, at a new Europe with a determinate content. Nietzsche responds to his own questions: "What is certain is that the same storm and stress tormented them and that they *sought* (*suchten*) in the same way these last great seekers (*Sucher*)!" It is a new way of seeking, a *Versuch* or experiment, not an end sought, that is significant. "Becoming one" means joyfully experimenting beyond the limits of nationality. These harbingers of the future are united only insofar as they model (often unknowingly) new forms of hybridity and cosmopolitanism. Nietzsche describes these "great discoverers in the realm of the sublime" in terms of multiplicity and variation: they are "born enemies of logic and straight lines, lusting after the foreign, the exotic, the tremendous, the crooked, the self-contradictory." Yet these excessive experimentalists "all broke and collapsed before the Christian cross," so their career trajectories fail as models of the good European; it is the mad, antinomian, and rebellious aspects of what Deleuze calls their lines of flight that offer a promise of the future. While these hybrids are all "Tantaluses of the will," failures who are unable to realize their (often unconscious) projects of transcending ethnicity and nationality, their experiments serve as initial models of living in the century of the multitude. Nietzsche has some fun at the end of the chapter's final aphorism, summing up Wagner's "*way to Rome*" as in his last music drama, *Parsifal.* He ends with a rhyme that asks of Wagner, "Is this still German?" and answers, "What you hear is Rome—*Rome's faith without the text.*" So Wagner used, again all unconsciously, the cloudy, nebulous spiritual geography of Germany, its mediating genius, to produce a hybridity contrary to his more explicit nationalist program. Wagner began, interculturally, with a music growing out of French romanticism, and ends with a displacement to Rome and Catholicism. He is a middle, a muddle, a cloudy milieu. If Nietzsche is appalled by Wagner's trajectory, he sees his transformations in the context of Europe's democratic movement, a movement of populations that produces new configurations of multiplicity, homogeneity, and hybridity, in the century of the multitude. But he has also shown that his early hero, the apparent arch-nationalist, is one more odd hybrid produced by Europe's nomadic fermentation.

Nietzsche returns, *it seems,* from the perspective of his "trans-European eye" to the local, the German, the place from which he began in *On the Advantage and Disadvantage of History* with his wish that the Germans could free themselves from the tyranny of history. Yet here the German essence is *aufgelöst* in the shifting tides of modern and modernizing Europe. What will be the direction of the earth? How can philosophy think the event now called globalization that Nietzsche first confronted in his attack on Hartmann's *Weltprozess?* Again, it will not be a question of either geopolitical strategies or establishing protocols for a new order of rank. The future of the earth cannot be known, planned, or predicted. Nietzsche's published remarks about such things as Russia's emerging power, the desirability of marriages between the Prussian officer class and wealthy Jews are best read as musings prompted by relatively immediate circumstances; they are no more important than his *Nachlass* note wishing that Germany should colonize Mexico.

Any attempt to extract or derive ideas like these from Nietzsche's writing would be a great waste of time, or possibly an amusing parlor game for initiates. What Nietzsche does do in the final chapter of *Beyond Good and Evil*, after taking the vivisectionist's knife to the virtues of his present and diagnosing the patient's complex ailments, is to end with a question: "What Is Noble?" Being noble or *vornehm,* of course, denotes a certain priority, whether natural or legal-traditional. We will proceed to interrogate two related aspects of Nietzsche's way of responding to the question, responses that don't provide conventional answers. The first has to do with time and its deformations. How can the *vornehm* live vigilantly with the challenge of a future that cannot be anticipated? How can they prepare themselves to be ready to seize the opportune moment by the forelock? This temporal openness must be further defended against the amortizing of the future in a system of debt, as will be shown in the *Genealogy* (billed as a clarification of *Beyond*). Second, nobility requires not only an understanding of the changing social structures of Europe (in Nietzsche' expanded sense of that term) but specifically of the ways in which the vigilant may be misled by the enthusiasms of the multitude, in a culture where various forms of celebrity or news of the day are confusedly taken as great events. How can we distinguish the merely spectacular flight of Zarathustra's simulacrum (or the media words and images commanded by a tyrant) from the true "great events" which come softly on doves' feet?

CHAPTER 4

Whose Time Is It?: *Kairos*, *Chronos*, Debt

There is a tide in the affairs of men
Which taken at the flood leads on to fortune;
Omitted, all the voyage of their life
Is bound in shallows and in miseries.

SHAKESPEARE, *Julius Caesar*

Perhaps genius is not rare at all: what is rare is the five hundred *hands* that it needs to tyrannize the *kairos*, "the right time," in order to seize hold of chance by the forelock!

NIETZSCHE, *Beyond Good and Evil*, 274

It is said that the world is in a state of bankruptcy, that the world owes the world more than the world can pay, and ought to go into chancery, and be sold.

RALPH WALDO EMERSON, "Gifts"

"WHAT IS NOBLE?"

"What is noble?" is the question Nietzsche poses with the title of *Beyond Good and Evil*'s final chapter. More precisely, "*Was ist vornehm?*" By the late nineteenth century, being *vornehm* has a range of meanings, including of course the older, more traditional sense of hereditary or monarchically conferred aristocracy. To be *vornehm* may also suggest being distinguished on the basis of achievement, character, and the like. Those who exhibit such nobility do so in virtue of their sustaining and furthering admirable causes and practices. In this vein, Nietzsche gladly accepted Georg Brandes's description of his ethos as "aristocratic radicalism." We may surmise that the noblest cause in this context is that of loyalty to the earth, its direction, and its future, the goals established in *BGE*'s yes-saying counterpart, *Thus Spoke Zarathustra*. *Treue*, of course, is a traditional aristocratic virtue. Nietzsche cites it as a distinctive virtue of the

Germans in the *Unmodern Observation* on Wagner. Zarathustra enjoins it in his initial set of speeches and reinforces its German associations in his talk "On the Thousand and One Goals."

To distinguish is to take first, to *vor-nehmen*. There is more than an auditory affinity between *nehmen* (taking or appropriating) and *namen* (naming), which is worth considering. It is fairly obvious in the older usage, where those who are officially *vornehm* have titles (duke, baron, and so on). They are not only first or primary, being selected before and above others, thus constituting a rank order, but they are given names that manifest their rank. Carl Schmitt examines the complicity of naming and appropriating in his philologico-political essay "*Nomos—Nahme—Name*." While not explicitly assigning any original etymological connection to the words, Schmitt points to the relationship between taking and naming in two examples that will raise suspicions. In much traditional European usage (although there is some variation) a husband who "takes" a wife "gives" her his name. More to the point for his purposes, "A land-appropriation is constituted only if the appropriator is able to give the land a name."[1] In keeping with Schmitt's admiration for the centuries of European conquest of the non-European earth, he celebrates the Spanish conquistadors who took the land of the Americas (oddly given the name of their cartographer Vespucci) "in the name of their Christian redeemer and his holy mother Mary." Yet the age of such naming seems past. Schmitt asks, "Has the power to name and give names disappeared? Has even what that means, what a name means, disappeared? . . . Where today are there still names?"[2]

The affinity of *nehmen* and *namen*, appropriating and naming, is also a Nietzschean theme, although it is not posed as directly as in Schmitt; more significantly, it is deployed for ends that subvert those of the jurist forever associated with the name "Nazi." The *Genealogy* speaks of "the seigneurial right (*Herrenrecht*) of giving names," which "even allows us to conceive of the origin of language itself as a manifestation of the power of the rulers (*Herrschenden*): they say 'this is so and so,' they set their seal on everything and every occurrence with a sound and thereby take possession of it (*nehmen es dadurch gleichsam*)" (*GM* I.2). Naming here is a right of the *vornehm*, those who name themselves as the good. In *Genesis*, of course, Adam is given dominion over the animals and so proceeds to name them. Naming is simultaneously a creation, privileged action, and expression of order of rank. Zarathustra had shown that his conception of nobility was not bound by tradition. In "Before Sunrise" he inverts a conventional understanding of naming and nobility as he confers a new title while claiming it as a restoration:

> For all things are baptized at the well of eternity and beyond good and evil; but good and evil are themselves mere intervening shadows and dampening sorrows and drifting clouds.
>
> Verily, a blessing it is and no blasphemy when I teach "Over all things stands the Heaven Accident, the Heaven Innocence, the Heaven Contingency, the Heaven Exuberance."
>
> "Lord Contingency" (*Von Ohngefähr*)—that is the oldest nobility (*Adel*) of the world, which I restored to all things when I redeemed them from their bondage under Purpose (*Z* III.4).

The oldest, most authentic nobility is Lord Chance. Nietzsche's naming here subverts all appeals to seniority, tradition, and origin, the attributes to which the "Vons" customarily appeal in justifying their preeminence. "What—or who—is noble?" In *Beyond*, the emphasis is more on the what than the who. In this chapter we will be concerned mainly with nobility as a mode of living one's temporality involving alert vigilance, freedom from the crowd's enthusiasms of the moment, and from the deadly deformation of lived time through economies of debt that mortgage the future. Yet toward the end of his meditation on nobility, Nietzsche will shift from a discourse of the "what" to one of the "who," as he seductively introduces the name of the tempter god Dionysus (*BGE* 295). In the new preface to *The Birth of Tragedy*, published in the same year as *Beyond*, Nietzsche explains the choice of that name in his first book. He had sought a counter-teaching to Christianity: "What was it to be called? As a philologist and man of words I baptized it, not without a certain liberty—for who knows the true name of the Antichrist?—by the name of a Greek god: I called it *Dionysus*" (*BT* P5). Nietzsche eventually claimed to speak in the name of the Antichrist. Later we will see that, both for him and Christian tradition, this is a highly political name, as can be seen already in its root, for Christ (*Χριστός*) is the anointed, the Messiah, who rules in the kingdom of God. It is also a name tied to the idea of an end-time, although in Nietzsche's version, it is an end of Christianity and the illusion of world-history, not the end of the earth.

TIMING IS EVERYTHING: *KAIROS* AND VIGILANCE

I cite an entire section from "What Is Noble?," which responds to the question by considering several different possible relations to those all-too-rare opportune moments that are often recognized regretfully only when they have passed.

> *The problem of those who wait.* Strokes of luck [*Glücksfälle*] and many incalculable factors are needed for a higher human, in whom the solution to a problem sleeps, to go into action at the right time—"into explosion," you might say. This does *not* usually happen, and in every corner of the earth people sit waiting, hardly knowing how much they are waiting, much less that they are waiting in vain. And every once in a while, the alarm call will come too late, the chance event that gives them "permission" to act,—just when the prime of youth and strength for action has already been depleted by sitting still. And how many people have realized in horror, just as they "jump up," that their limbs have gone to sleep and their spirit is already too heavy! "It's too late"—they say, having lost faith in themselves and being useless from that point on.—What if in the realm of genius, the "Raphael without hands" (taking that phrase in the broadest sense) is not the exception but, perhaps, the rule? Perhaps genius is not rare at all: what is rare is the five hundred *hands* that it needs to tyrannize the *kairos*, "the right time," in order to seize hold of chance by the forelock! [*um den Zufall am Schopf zu fassen!*] (*Beyond Good and Evil*, 274)[3]

Let us read this passage carefully, in the context of Nietzsche's *Beyond Good and Evil*, in order to deepen his conception of time and political change. More precisely, let us consider the question of timing in relation to Nietzsche's political categories of multitude, nomad, and hybridity, which modify or supersede more traditional ideas of people, nation, and state; these categories take us beyond the "Peoples and Fatherlands" that form the ostensible theme of the eighth chapter of *Beyond Good and Evil*. This reading aims at articulating his suggestions about the *kairos*, or right time, with what he says more specifically about *his* time, which he calls "the century of the *multitude* [*Menge*]*!*" (*BGE* 256). In *BGE* Nietzsche is thinking about what it means to have a philosophy of futurity, that is, of the unexpected, unpredictable, incalculable event. Further, he provokes us to ask what it is to think in the time of the multitude, a diverse collection or assemblage of hybrid, nomadic human types who are culturally attuned to spectacle, to theater in the widest sense. The task of those who would be vigilantly alert for opportunities of significant political change in such a context depends on their comprehending the character of their time, a time called "the century," and on their ability to avoid the seductions of theater and political theater to which the multitude is liable. To clarify this line of thinking, it will be necessary to follow Nietzsche more closely than most of his readers have done in distinguishing the multitude from some of his other categories of social and political analysis, such as masses, herd, or rabble.

In Nietzsche's published and unpublished writings, the term *kairos* appears only once (according to the digitalized version of his works from de Gruyter). As the aphorism on "The problem of those who wait" indicates—and as we will soon confirm—this word is not easy to translate. Its provenance is both classical and Biblical. The *kairos* is, roughly, the right time, significant moment, turning point, or unexpected and unique hinge of opportunity. It is a "stroke of luck," a serendipitous moment, incalculable, unpredictable—it partakes of the character of the event in some of the senses given this term by thinkers like Heidegger, Deleuze, Derrida, and Badiou (a critical history of thought from Heidegger to Badiou could revolve around explicating some of the senses and implications of "event"). Since 2001, two book-length English language commentaries have appeared on *Beyond Good and Evil: Prelude to a Philosophy of the Future.* Yet neither of them mentions this expression, and neither treats the aphorism in which it appears in any depth.[4] Perhaps *BGE* is the least understood, the least well read of Nietzsche's major texts. This prelude to a philosophy of the future still has a future. The theme of *kairos*, the opportune moment, would seem to be central to any philosophy of futurity, of the *Zu-kunft*, of that which arrives, arises, or emerges.

Given that futurity is precisely that which cannot be calculated or intentionally produced, in the perspective of a "*Philosophie der Zukunft*," speculating about any specific character or content the future might have is a chancy business.[5] Nevertheless, as I will be suggesting, Nietzsche's call for loyalty to the earth and his aspiration to develop a "philosophy of the Antichrist" provide some pointers. Nietzsche's project, articulated most explicitly in *Beyond*, is to think, not the content of the future, but futurity itself, and to provoke his readers to a certain vigilance in their time and place. "The problem of those who wait" is our problem in so far as we are moving deeper into the questioning indicated by the title of *Beyond*'s final chapter, "What Is Noble?" in which the aphorism on waiting appears. In other words, it is (first) a problem that concerns our understanding and evaluation of those who wait among the other types, roles, and characters that are being assessed in terms of the possibility of nobility. Yet it is also the problem posed *to* those who wait, a problem of which they may not even be aware.

The aphorism encourages us to formulate the problem of waiting, a problem which is also necessarily ours as thinkers of futurity. Some "wait" completely unconsciously, waiting only in the sense that they have a set of skills, instincts, aptitudes, and thoughts that could respond creatively and responsibly to the stroke of luck, the quick passage of a fortunate chance, the *kairos*. Some sense with varying degrees of awareness what might be possible, yet the event in the strong sense tests and transforms the limits previously given to the possible. Articulating the problem of those who wait, articulating it *as* a problem, is itself a form of vigilance, a watchful, self-aware, questioning waiting. Indeed, what is a problem? Is it a question? Or, as Deleuze suggests, does a philosophical problem have the form of a complex of questions that are multiply linked and associated, but that exhibit no unique internal order or hierarchy?[6] A problem, then, calls for a judgment that either reduces the problem to an ordered series of questions (like Kant's determinate judgment) or provides some less structured, more pluralizing way of exploring the complexity (variations on Kant's reflective judgment that go by the names of hermeneutics, genealogy, deconstruction, and the like).

Whether Nietzsche's chapter on the question of nobility answers the question or multiplies and deepens it is not obvious. Some readers, supposing that it promises a certain kind of answer, have judged the chapter and the book to be incomplete; some say that the concluding aphorisms are so scattered that "things fall apart."[7] Yet to take the text at face value, to read rather literally, we should ask whether in this unique reference to *kairos*, Nietzsche may be encouraging us to think together the questions of futurity and nobility. *Does nobility then involve a certain relation to futurity?* In this case, "the problem of those who wait" would be central to its intent. Nobility, it must be remembered, is not only an individual character trait, but a form of social and political distinction, even when decoupled from ideas of hereditary aristocracy.

What is the right time, the *kairos*? How can we recognize it and be prepared for it? To paraphrase Meno's challenge to Socrates, how will we know it when we see it? And how can we search for it when we don't know what it is? To these we must add a temporal dimension that Plato neglected when he turned the issue into one of *anamnesis*: how can we recognize, catch it, and respond to it at the right time? How can we be worthy of the event? The Greeks acknowledged this problem in their personification and images of *kairos*; he became a figure with locks of hair above his face, but bald in back. Images of this figure are found in antiquity and in early modern emblem books.[8] *Kairos*'s large upper wings are matched by a smaller pair on his ankles, suggesting the swiftness of his passage. He appears quickly and unpredictably. In a moment he has flown by. Some images show him holding or examining a scale that is out of balance, indicating that the moment tilts toward possibilities that can be realized by the alert agent seizing the time—what Machiavelli called *occasio* or opportunity as distinguished from *fortuna* or chance.[9] *Kairos* must be seized by the forelock as soon as he appears, in a face-to-face encounter; you can't hesitate until he has passed and the chance is lost. The figure is evoked in *Zarathustra*'s chapter "On Redemption." Zarathustra prepares to cross a great bridge, where he is surrounded by a crowd of cripples and beggars. (The scene is reminiscent of the Gospels' stories of Jesus's preaching and magical healing.) A hunchback from this assemblage addresses him:

> "Behold, Zarathustra! Even the people are learning from you and coming to believe in your teaching: but in order for them to believe you completely one thing more is needed—you must first persuade us cripples too! Now here you have a fine selection, and verily, an opportunity with more than one forelock! (*eine Gelegenheit mit mehr als Einem Schopfe!)*" (*Z* II.20).

Nietzsche clearly alludes to the stories of Jesus's magical healings and exorcisms, which occupy more space in the Gospels than do his teachings and which are powerful tools in winning the affections of the multitude (a term whose Biblical associations Nietzsche exploits, as we will soon see).[10] Zarathustra declines the opportunity to grasp these opportunities, and his speech "On Redemption" explains that taking away the hunchback's hump would be to injure his spirit. Yes, broken and fragmented bodies and spirits abound, but the only genuine redemption would be one that freed all humans—not only the obviously disabled—from their resentments against "time and its 'it was.'" Seeking a different kind of redemption than does Jesus, Zarathustra looks for opportunities that could transform our experience of earthly time, rather than escaping from it or reversing its marks.

The future cannot be known by prediction or prophecy, but precisely, as Derrida reminds us, it is that which is *à venir*, a *Zu-kunft* that has yet to manifest itself. Therefore it is not a future that we plan, install, or force, but one assembling itself through "happy accidents" and "incalculable circumstances." *Beyond Good and Evil* names the site of arrival, the place from which the *kairos* might become evident, might be seized by the forelock, and possibly even tyrannized. Yet since great events slip by almost everybody, even those with the resources for greeting them, those who might have been equal to the event, are almost always too late.

What is the problem of those who wait? The phrase can be read in at least two ways. In one sense, the emphasis would be on the situation of waiting itself; in another, it would be a question of what the problem is that finds them waiting for a solution. Nietzsche's explicit concern in the aphorism seems to be with the first: what does it mean to wait for an opportunity, a happy chance, that never comes? What of those who see it come too late, when waiting itself has exhausted their "best youth and power"? These are perhaps the unhappiest of those who wait, for they see the moment slipping through their hands. And what should we say of those who, having the power to seize the *kairos*, remain unaware that this is their vocation, and consequently allow it to elude them? Nietzsche's reference to this last possibility indicates that from his perspective there is a kind of objective waiting, a waiting of which one can be unaware. Such "Raphaels without hands" do not know that they are capable of rising to the occasion, or that there could be an occasion that would call forth their powers.

The *kairos* of this aphorism should be understood as part of a system of terms including events, great events, and great noon. Events, that is, genuine events, as Nietzsche says both here and elsewhere, are rare and unpredictable (cf. chapter 1). Yet we should not hear Nietzsche's frequently hyperbolic rhetoric as implying that only the rarest of types, for example, the *Übermensch*, are capable of recognizing, yielding to, and ultimately tyrannizing the event. The events in question are much rarer, according to *BGE* 274, than are those who might catch and tyrannize them. Indeed, this passage should be read alongside the much-quoted lines in *Schopenhauer as Educator*, where Nietzsche says that in principle all people could look beyond themselves to discover an exemplar or educator, a point emphasized by Stanley Cavell (*UO III.* 6).[11] This theme is explicit in the passage from Lessing's *Emilia Galotti*, the source of Nietzsche's phrase "Raphael without hands." There the Prince of a small Italian renaissance state has become infatuated with Emilia, the daughter of a rival, who is promised to another. The painter Conti appears at the play's beginning and shows the Prince two paintings, first of his last month's love and then of Emilia. When the Prince heaps extravagant praise on the painting, the artist replies with this speech:

> "Nevertheless, this picture still left me very dissatisfied with myself. And yet, on the other hand, I am very satisfied with my dissatisfaction with myself. Ha! What a pity that we do not paint directly with our eyes! How much is lost on the long path from the eye, through the arm, into the brush! But the moment I say I know what has been lost here and how it has been lost, and why it had to be lost: I am just as proud of that, in fact prouder, than I am of what I did not allow to get lost. Because from the former I recognize more than from the latter that I am really a great artist, but that my hand isn't always. Or do you think, Prince, that Raphael would not have been the greatest artistic genius had he unfortunately been born without hands? Is that what you think, Prince?"[12]

What can it mean, then, to *tyrannize* the *kairos*, to actualize the power of the five hundred hands of those who wait? Earlier in *BGE*, Nietzsche characterizes the Stoics as spiritual tyrants who insisted on seeing nature as Stoic:

> Some abysmal piece of arrogance finally gives you the madhouse hope that because you know how to tyrannize yourselves—Stoicism is self-tyranny—, nature lets itself be tyrannized as well: because isn't the Stoic a *piece* of nature?
>
> . . . what happened back then with the Stoics still happens today, just as soon as a philosophy starts believing in itself. It always creates the world in its own image, it cannot do otherwise; philosophy is this tyrannical drive itself, the most spiritual will to power, to the "creation of the world," to the *causa prima*. (*BGE* 9)

Stoic tyranny involves the thinker forgetting or denying that she has imposed a meaning on the world; the thinker claims to discover what she has herself constructed. In his comment on the Stoics, Nietzsche admires the decisive energy of such tyranny while resisting its temptations. But is this what he means by tyrannizing the *kairos*?

Part of the problem of those who wait is understanding what kind of tyranny is possible with regard to the *kairos*. What is the relation of tyranny and time? This tyranny will *not* be a capture or seizure. In *BGE* 279, we hear that sad people are prone to grasping and clutching at their *Glück*, a response that Nietzsche's worldly wisdom realizes will only drive it away. This would be an unsuccessful tyranny, as Brecht reminds us in *Der Dreigroschenoper*: those who chase after *Glück* will always find that it runs away faster than they can pursue it. Therefore Nietzsche speaks of *Glücksfälle*, happy accidents. What is *Glück*? As *BGE* 274 suggests, it is not a state or condition or experience that can be held, exchanged, or distributed, but an emergent event, so the word itself is appropriately translated as "happiness," "fortune," or "luck." The history of the English word "happiness" provides an instructive parallel. It derives ultimately from the somewhat archaic "hap," used to describe in the most general way the occurrence of any sort of event; "happen," "perhaps," "mishap," and similar words are derived from it. In recent language, events "happen," in one of the few remnants of the middle voice in English. To be happy was to experience a fortunate happening. Only in later usage does "happiness" become something to be deliberately pursued, possessed, grasped, or held as a treasure. One of Nietzsche's translators has—happily—translated *Glücksfälle* as "serendipities," a word that owes its origin to the supposed wonderful chance discovery by European explorers of the marvelous island of Serendip (Ceylon or Sri Lanka), which could remind us of *Zarathustra*'s *glückselige Inseln* (*Z* II.2–18).

Much of Nietzsche's thinking and questioning in "What Is Noble?" is apotropaic, warding off the premature embrace of the all too timely, on the one hand, and, on the other, resisting the resigned skepticism that blinds itself to that which arrives. It is not so much inconclusive as it is a complex set of warnings against premature conclusions. It is directed at keeping the five hundred hands as primed as they can be. What is nobility now, for those who wait? Among other things, it is avoiding premature, precipitate action and knowing how to avoid such temptations. In *BGE* 277, Nietzsche reminds us of the classic paradox that once you've built your ideal house you have already thought beyond it to what it should have been. We should learn not to be overeager in anticipating the moment. We are also reminded of the necessity of masks and disguises, to protect us in our condition of vigilance from those who would interfere if they knew our vocation; so we are advised on the value of feigning stupidity through enthusiasm (288) and of the strategic necessity of writing in order to hide our thoughts (289). This advice can be read as an amplification of the aphorism that precedes and introduces 274, where Nietzsche names the *kairos*. The one who strives for great things will see others as obstacles, stepping stones, or resting places—even when she becomes involved with them, she remains solitary at a deep level (273).

The apotropaic dimension of *Beyond Good and Evil* gains a more specific focus by considering another form of temporality informing the situation of those who, wittingly or not, are waiting for a *kairos*. This is the character of the time in which Nietzsche and his readers live. It is not a privileged moment but a stretch of time, a manifestation of modernity, or, to take the classical contrast with *kairos*, it is *chronos* (roughly duration, discussed further in what follows). To think of futurity philosophically, it is necessary to think not only time as event and *kairos*, but time as our time, the time in which we are waiting.

So it should be possible to articulate what Nietzsche has to say about the *kairos* with his analysis of his own time. Specifically, let us look again at the apparently definitive statement "this is the century of *the multitude* [*Menge]!*" (*BGE* 256). Coming at the end of the chapter "Peoples and Fatherlands," this declaration asks to be taken seriously as an emphatic judgment about the age of nationalism under analysis.

"Peoples and Fatherlands" addresses a number of questions having to do with states, nations, and the future of Europe. (Europe is understood here as a cultural rather than a strictly geographical entity, including all peoples whose cultures and polities are strongly indebted to a complex of social, artistic, and political practices that arose in geographical Europe in modern times, that is, roughly from the time of the scientific revolution and the establishment of independent national states in the seventeenth century). The definitive claim—"this is the century of *the multitude* [*Menge*]!"—appearing as an interjection set off by dashes, could be said to be the chapter's conclusion, implying as it does that we must think *beyond* peoples and fatherlands in order to understand the emerging political contours of modernity.

This thesis comes at the end of a chapter ostensibly devoted to "peoples and fatherlands"—specifically Germany, France, England, the Jews, and others—yet it undermines or significantly qualifies the assumption, with which a reader might have begun, that Europe consists essentially of nation-states on which its future and its politics depend. That assumption is also put into question by the repeated observation that "Europe wants to be one" (256); that claim has attracted some attention, typically in connection with Nietzsche's references to "the good European." However, it has not been sufficiently noted that Nietzsche is *reporting* this desire; if he endorses it, it is not obvious why he does so. He does not see the desire as leading immediately or directly to social homogeneity (which Nietzsche tends to designate with words like "masses [*Massen*]" or "herd [*Heerde*]") but to new forms of diversity and multiplicity, that is, to the formation of a multitude. The multitude arises because Europeans have become increasingly nomadic and detached from their origins or ancestry. This must eventually weaken the power of the nation-state and its claims to legitimacy (*HAH* 472, 475). In the process, as Nietzsche illustrates by discussing a number of exemplary cultural figures from Napoleon to Wagner, these contemporary Europeans take on hybrid forms, creating new mixtures drawn eclectically from a variety of traditions. They develop *beyond* peoples and fatherlands. It is the multitude's century, not the nation-state's, and it is also not the century of the masses. The multitude's roots in national identities have been significantly loosened, and they have become an audience for those leaders of the near future whom Nietzsche describes as tyrants, including the most spiritual sort (242). Indeed, it is a condition of the appeal of these tyrants that they can compete for the allegiance of a multitude. This account recognizes that the politics of the era of the multitude will have a strong dimension of theater and spectacle.

Philosophical commentators tend to neglect Nietzsche's formulation. Consequently, they miss some important suggestions about what it means to articulate a philosophy of the future and of futurity. Consider a representative critical remark at the conclusion of an otherwise insightful article on some of Nietzsche's complex position(s) on race. Yirmiyahu Yovel says that "there is a marked lacuna in [Nietzsche's] thinking—the lack of a positive philosophy of the 'multitude.' Politics is not about the happy few, but about those ordinary people, the modern mass or 'herd' which Nietzsche did not care about and did not make the topic of any positive philosophical reflection." Yovel goes on to say that this political lacuna left (and still leaves) Nietzsche open to abuse by fascists, Nazis, and the like.[13] Here Yovel, like many other readers, assumes that the terms "multitude," "mass," and "herd" are identical. Whether Nietzsche has a "positive philosophy" of the multitude—and what this would involve—remains to be seen, but we must begin by reading his words. The chief elements in the makeup of the herd appear to be lack (not loss) of individuality and unthought, unquestioning submission to the herd's shepherd or acceptance of its collective, instinctive behavior. The terms "mass" and "masses" suggest great numbers and homogeneity, and one can often hear Nietzsche attempting to transvalue what he thought socialists were saying about the masses when he uses this term. The *multitude* of *BGE*, however, is composed of diverse individuals and is strongly inflected by many hybridic and nomadic strains (cf. 242, 256). Another obvious sign that Nietzsche's thesis has been neglected or misunderstood is found in the inconsistencies in English translations. The Cambridge translation renders "*Menge*" in *BGE* 256 as "masses," following the lead of the century-old Zimmern version. Masses, however, are almost always taken to be homogeneous in Nietzsche (and in many other nineteenth-century writers), lacking the diversity of the *Menge*. Walter Kaufmann's "crowd" is somewhat better, especially because an audience can be spoken of as a crowd. *Menge* appears in a number of other contexts in *BGE*, and the translations tend to be inconsistent in their renderings. So the *Menge* not only designates a multitude but oddly gives rise to a multitude of meanings. Deeper perhaps than the problem of linguistic translation is a continuing practice of reading Nietzsche as an exclusively aphoristic writer whose books are without structure. More recently, readers have become alert to Nietzsche's development of systematic arguments in specific books, especially in the case of the *Genealogy of Morality*.[14] Nietzsche insisted that his writings be approached through slow and careful reading (*GM* Preface). More specifically, in *GM* I he undertakes a discriminating, differentiating look at the terms used to name human groups or types. Nietzsche advertised *GM* as an expansion and clarification of *BGE* (cf. *KSA* 14.377). There Nietzsche asks us to pay attention to distinctions, even subtle nuances, in the oldest Greek and Latin terms that the masters and slaves use to describe one another. He notes the nuances of tenderness or compassion in some of the nobles' names for the slaves, urging us to hear "the almost kindly nuances which the Greek nobility, for example, places in all words that it uses to distinguish itself from the more lowly people [*das niedere Volk*]" (*GM* I.10). Nietzsche reinforces the methodological point, proposing that some learned academy invite the submission of essays on the question of how linguistics illuminates moral concepts (*GM* I.17). While we can be grateful for the work of scholars like Mauss, Benveniste, and Foucault, who have made exemplary genealogical discoveries in this spirit, we should also apply it to the reading of *BGE* for which *GM* is said to serve as a clarification.

It seems clear that *Pöbel* (rabble), *Heerde,* and *Massen* are always employed with a tone of contempt in *BGE* and elsewhere. Nietzsche is appalled by the possibility of the formation of a strong, uniform herd; disdainful of the rabble, he sees the homogeneity of the masses stifling the rise of noble individuals. History written from the standpoint of the masses will derive laws from a statistical study of "the lowest mud- and clay-strata of society . . . but are the effects of inertia, stupidity, love, and hunger to be called laws?" (*UO* II.9). There is little or no contempt when he speaks of the *Menge*. Accordingly, I seek to convey a more neutral tone, following Grimm in linking the term to the Latin *multitudo*, translating it with the English *multitude* (we might also think of words like "throng").[15] I take the dominant note here to be multiplicity, a multiplicity that does not (or does not *necessarily*) denote uniformity.[16] Elsewhere Nietzsche explicitly distinguishes *Menge* and *Masse* in terms of the greater diversity of *Menge*. In *Gay Science* he says that in Greece "there must have been a multitude of diverse individuals [*eine Menge verschiedenartige Individuen*]," contrasting this, later in the aphorism, with the homogeneity of the *Masse* (*GS* 149).[17] In this aphorism Nietzsche makes a very clear distinction between *Masse* and *Menge*. The topic is "The failure of reformations"; Nietzsche asks why Luther, whom he generally describes as a vulgar peasant, was able to accomplish a reformation in northern Europe when much more gifted spirits like Pythagoras, Empedocles, and Plato failed. He concludes that "every time the reformation of an entire people fails and only sects raise their heads, one may conclude that the people is already very heterogeneous [*vielartig*] and is starting to break away from crude herd instincts and the morality of custom [*Sittlichkeit der Sitte*]."

BGE 256 develops this thought about the multitude by examining the careers of exemplary cultural figures whose hybridity and internal multiplicity reflect both the heterogeneity of the *Menge* that idolizes them (cf. 269) and the European *Menge*'s desire to be one; but just as Greek reformations failed, so such unification is unlikely so long as the population remains diverse. In "Peoples and Fatherlands" Nietzsche discusses both factors that could encourage unification (such as the slow generation of adaptable supranational and nomadic types [242]) and the actual diversity that leads not to homogeneity but to varied forms of hybridity. These artists and political figures, whose achievements arise from mixing and synthesizing novel combinations of various cultural traditions, resemble one another in the form but not the content of their hybridity (e.g., Heine's German-Jewish persona is distinct from Stendhal's

Franco-Italian one). The *Menge*, it seems, is like these hybrid cases insofar as its members too tend to be of mixed but not uniform heritage.

As the context of *BGE* 256 suggests, the *Menge* is, among other things, an audience. From the beginning of the aphorism, we are in the world of theater, as Nietzsche explains that the nationalistic politics of the day is "a politics of dissolution [*auseinanderlösende Politik*]," which necessarily can only be a politics of the theatrical interlude (*Zwischenakts-Politik*). They are those listening to such cultural stars as Wagner, Stendhal, and Heine; these and others first taught their century the concept "higher human." The chapter begins and ends with attempts to assess Wagner in terms of his presumed exemplification of German national culture; in each case the supposed identity is considerably loosened under examination. He is to be understood as a performer in his relation to his audience, the multitude or throng. Given this theatrical context, we should recall that the *Menge* plays a prominent role in Goethe's *Faust*, where it is the subject of a dialogue that frames the play. Goethe is one of Nietzsche's constant touchstones and is included in his list of the century's unconscious seekers for "the soul of Europe, the one Europe." In Goethe's "Prelude in the Theater," the *Menge* is the audience. As the dialogue of director, poet, and clown shows, the *Menge* is not a universal class, but rather those attending the play; there is no indication that they are a representative sample of the people at large. The director wants to please them, especially because they "live and let live." They want a surprise, and "even if their taste is not the very best, they have read quite a bit." This suggests a semi-educated but reachable audience. The director's long description of the *Menge* is worth quoting because it emphasizes their variation in mood, attention, and interest, as well as implicitly acknowledging that they are all likely to be comfortably situated in social and economic terms:

> Do not forget for whom you write!
> They come when they are bored at night,
> Or gorged on roast and relish, spice and capers,
> And—this is the most wretched plight—
> Some come right after having read the papers.
> They come to us distracted, as to a masquerade,
> Propelled by nothing but curiosity;
> Their dresses and their jewels, the ladies would parade,
> And act without a salary . . .
> One half is cold, one half is raw.
> After the play, one hopes to play at cards,
> Another for an orgy in a harlot's bed.[18]

Later in this dialogue, the clown emphasizes other internal differences in the *Menge*, especially between the young and old.

While the poet speaks disdainfully of the *Menge*, the clown and director ridicule his interest in perfection and the judgment of posterity. To think of this as the century of the *Menge*, then, is to see this as a time in which power is obtained through spectacle presented to a variegated audience. Now, appeals to "peoples and fatherlands"—national identity or traditional patriotism involving essentialist or autochthonist claims—are not likely to be effective, except insofar as these are ingredients of a successful spectacle. While Nietzsche observes that all of the exemplary hybrid cultural figures he cites as the time's first educators with respect to the idea of higher men had severe limits (e.g., all relapsed into Christianity), they did delineate the structure of the emerging political arena.

Nietzsche's references to the multitude in *BGE*, while never admiring, are rather nuanced. Consider *BGE* 213, where the philosopher is said to have been born and bred for a "higher world" and enjoys the "sovereignty of his ruling gazes and downward gazes"; accordingly, he feels his "separation from the multitude with its duties and virtues." Nothing derogatory is said here about the multitude, with their own duties and virtues. The philosopher is also separated from everybody else. If the philosopher were simply said to be separated from the rabble, the masses, or the herd, all of whom are typically described with disdain, the separation would not be as definitive as Nietzsche wants to make it.[19]

That the *Menge* is not a universal class of all human beings, or all those within a certain territory or political unit, is evident from a discussion of their reverence for "great men" (269). Here the multitude is again understood as an audience, one that often admires unwisely, but is distinguished from a more universal class. This admiration is typically naïve; in contrast, "the psychologist" is aware of the pitiful shortcomings of the figures generally considered to be great. The psychologist—a role Nietzsche plays when he analyzes the "higher humans" (as in 256)—suffers from observing their admiration: "Perhaps the paradox of his condition becomes so horrible that the multitude, the educated, the enthusiasts [*die Menge, die Gebildeten, die Schwärmer*] develop a profound admiration for the very things he has learned to regard with profound pity and contempt . . . ." Nietzsche takes this contemporary phenomenon as a clue to "what has happened in all great cases so far: the multitude worshiped a god,—and that 'god' was only a poor sacrificial animal!" The apposition of "multitude, educated, enthusiasts" indicates the relative selectivity in the concept of multitude, as opposed to herd and masses. They are those with sufficient interest and motivation, whatever their other differences, to care intensely about "great men." While such things may always have happened with the multitude and the objects of its admiration, we are now living in the very time of the multitude, their century.

Nietzsche's terminology and psychological analysis may draw on what he called the greatest German book, Luther's Bible. In the gospels, and especially in Mark, Jesus is often surrounded or followed by a curious, fascinated, and admiring multitude, a multitude whose enthusiasms are fickle and can melt away in times of trouble. The rough correspondence of Nietzschean and Biblical social analysis becomes increasingly evident when we realize that in *Zarathustra* and the Gospels, three social groups are distinguished in relation to a single charismatic figure: disciples, the multitude, and the people generally. As a contemporary Biblical scholar phrases it, "disciples/crowds/whoever."[20]

One more surprising distinction is evident when Nietzsche describes women's attitude toward the presumed great men. While the multitude adores its "great men" without qualification, not everyone does.

> It is easy to imagine that they [higher humans] will soon be subject to eruptions of boundless and most devoted pity from women in particular (who are clairvoyant in the world of suffering and whose desire to help and save far exceed their ability to actually do so). The multitude, the adoring multitude, above all, do not understand this pity . . . .[21]

Recently, Antonio Negri, Michael Hardt, Etienne Balibar, and others have spoken about the multitude as a social, political, and revolutionary agent. Nietzsche's use of the concept should be distinguished from these contemporary versions, and from Spinoza's, which is typically cited as their precedent. Spinoza's multitude are the people at large, the many, understood as a combination of individuals that can act collectively; Balibar emphasizes the potential for affective community, which enables such action.[22] This conception of the multitude is much more inclusive than the theatrical one, as Nietzsche conceived it, and specifically attributes a form of agency to the multitude. However, one might think that much of the global audience that follows international music, soccer and the Olympics, film, and the spectacles of war and terrorism could be described as a contemporary version of the *Menge*, indeed as a multi- and intercultural expansion of it. Like Nietzsche's *Menge*, it does not assume uniformity among the multitude, but preserves the strong possibility of differentiation. For Spinoza, the multitude is not as passive as the later thinker deemed the herd, the masses, and the rabble. Hardt and Negri claim to follow

Spinoza in *Multitude: War and Democracy in the Age of Empire.* They see the multitude as a collective agent composed of diverse individuals in a world of hybridity and nomadism. In contrast to the multitude, they understand the masses as a rather dated concept, tied to the industrial economy that dominated classical Marxist thinking. While Nietzsche does not look forward to the *Menge* taking political power, insofar as they form an audience, it is their applause or boos, favor or disfavor, that determine which would-be leaders hold power and for how long. Those who want to deploy Nietzsche's thought in behalf of an agonistic democracy might find this combination of the political and the theatrical of mixed value.[23] The multitude has some power here, but it is understood in terms of its response to competing spectacles. Perhaps the "tyrants of all sorts, including the most spiritual" that Nietzsche sees emerging include those who stage media spectacles and even own or control the media; they might range from media powers, moneyed interests, and politicians waging television campaigns, to terrorists staging attacks like that of September 11, 2001.

Why have Nietzsche's readers ignored the distinctive qualities of the multitude? The frequent conflation of multitude, herd, and mass assumes that Nietzsche sees Europe on the way to producing a uniform, homogeneous, heteronomous population—if it is not already there. However, his position is more complex. The multitude is formed by a mixing of races, cultures, ethnicities, and so on. This might result eventually in the formation of herds and masses, but it need not. Exemplary here is Nietzsche's discussion of the emergence of what we think of as the Greeks from a mixing of Mongols, Semites, and others (*KSA* 8.96).[24] Mixing was the necessary precondition for creating the Greeks. Neither they nor any other group has a simple or pure origin. In *BGE*, Nietzsche looks forward to the possibility of something analogous issuing from Europe's contemporary mixing. If Europe wants to "become one," if this is the form of its desire, the "truth" or upshot of this desire (as Hegel would say) is that it becomes a great mixing bowl that could generate a new ruling caste. While this thought can be and has been abused, it is quite distinct from any view about the inevitable dominance of the uniform herd. In this respect it is compatible with much recent political thought focused on questions having to do with the movement and mixing of peoples, the formation of new cultural configurations, and the constitution of a diverse population. Nietzsche's terms for these phenomena are nomadism, hybridity, and multitude.

The multitude, like the throng Zarathustra finds in the marketplace, are enthralled with the spectacle of the moment; they are likely to be fascinated with a rope-dancer or give their ears to the fire-hound, who barks and howls about political ephemera. *Grosse Menschen*, if there were any, would be associated with great events, in the manner of Hegel's world-historical figures. Nietzsche approaches the situation of the multitude through the figure of the psychologist, who studies "the more select cases," those often taken to be *grosse Menschen*. The psychologist discovers that "it seems to be the rule that higher people come to ruin." He is tormented by repeatedly uncovering the "eternal and all-encompassing 'Too-late'" of these higher people, anticipating what will be said just a few aphorisms later about those who see the *kairos* only after it has flown by. The psychologist must resist the temptation of pity—precisely the situation of Zarathustra with the higher humans. At this point he finds himself in opposition to the *Menge*:

> The paradox of his situation may even reach the frightful point where those cases that have triggered in him great pity as well as great contempt, have triggered in the multitude, the educated, the enthusiasts, a feeling of great reverence; theirs is a reverence for "great humans" and performing animals, for whose sake we bless and esteem the fatherland, the earth, the dignity of humanity, and ourselves; those whom we ask our children to look up to and to emulate (269).

Note that Nietzsche has silently enlisted the pity of his reader for the psychologist, so that we find ourselves in (or resisting) a situation parallel to his. The misplaced reverence of the crowd prevents them from detecting genuinely great people or events. The melancholy of the psychologist threatens an equivalent oblivion regarding the future. As noted earlier, "the multitude, the educated, the enthusiasts" are *not* the herd, masses, or rabble. These noisy and exaggerated enthusiasts fail to see that their stars are not so different from performing animals. They see the great humans as justifying the earth, the fatherland, and their own dignity.

Perhaps, Nietzsche continues, it has always been so with the multitude, adoring an imagined "god" who was "only a poor sacrificial animal." The great humans themselves are woefully unprepared for the *kairos* because they are "precipitous in their trust and distrust," "people of the moment," and likely to be swayed by "intoxicated flatterers." Nietzsche's sample list of such supposed great ones—Byron, Musset, Poe, Leopardi, Kleist, Gogol—overlaps with the one provided in *BGE* 256. Such figures often become the target of women's "limitless, utterly devoted *pity*." Here Nietzsche draws a distinction between women and the general, reverent crowd, which admires rather than pities. Perhaps the point is to introduce a revaluation of what Christianity—for Nietzsche a feminized religion—saw as the distinctive *kairos*, the crucifixion and resurrection of Jesus. Indeed, most of the discourse on *kairos* for the last two thousand years, beginning with Paul's letters, has centered on this presumed messianic moment (the pre-Christian sense of the theme tended to reappear under names like *Fortuna* and *Occasione*). Nietzsche refers to this as a "holy tale and camouflage." Perhaps Jesus, whose demands for love knew no limits, and had to be disappointed by all human love, had to become a martyr for a God he constructed as all love. Nietzsche would then be deconstructing the Christian *kairos* by seeing it as a misunderstanding born of enthusiasm, pity, and a misunderstanding of love.

At the end of this aphorism Nietzsche reflects, asking us to acknowledge our own involvement, perhaps our pity, first for the unnamed psychologist and then for Jesus: "But why should we muse on [or give ourselves up to (*nachhängen*)] such painful things? Assuming that we are not compelled to do so." But are we not compelled to think about such "painful things" insofar as we are becoming thinkers of futurity? Is it not the promiscuous identification of great events and false *kairoi*, surrounding us in the century of the multitude, that compels such musing on these painful things? Is this not a major dimension of the problem of waiting in our time—to avoid the simulacra of great events that are the daily work of the society of the theater and the spectacle?

In "Peoples and Fatherlands" Nietzsche describes our age of democratization—that is, of the multitude, with its increasingly nomadic, cosmopolitan, and hybrid population—as "an involuntary exercise in the breeding of tyrants—understanding that word in every sense, including the most spiritual" (242). Is "our time" then helping to breed spiritual tyrants who can recognize and tyrannize the *kairos*? This question calls for thought about both time and tyranny. What do we mean by time here? *Kairos* is typically thought in relation and in contrast to *chronos*, that is, to a progressive, devouring time, what John Locke called "perpetual perishing." *Chronos* typically extends into such measures as days, years, and centuries. When Nietzsche says that "this is the century of the *multitude*!" I take him to be speaking the language of *chronos*, indicating a relatively extended stretch of continuous time (it is *chronos* that lends itself to the spatialization of time that Bergson and others subject to critique). He is not suggesting that this is a span of exactly one hundred solar calendar years, but naming an epoch that could be the context of *kairos*. As the Grimm *Wörterbuch* confirms, *Jahrhundert* is a translation of the Latin *saeculum*; the latter means something like an age or an epoch, the span of the longest human life. Thus a *saeculum* or *Jahrhundert* is finished when the last people to have experienced that era have died. It is the time of "living memory."

Giorgio Agamben cites a telling definition of *kairos* and *chronos* in his study of Paul, *The Time That Remains*:

> In general, *kairos* and *chronos* are opposed or heterogeneous, which is certainly true. But decisive here is not simply the opposition, but the relationship between them. What do we have when we have a *kairos*, an occasion? The most beautiful definition of *kairos* I have ever found is in the *Corpus Hippocraticum*, and it is one which in fact characterizes *kairos* with respect to *chronos*. I will quote this definition: *chronos esti en ho kairos kai kairos esti en ho ou pollos chronos*, "the *chronos* is where we have *kairos* and the *kairos* is where we have a little *chronos*." Mark the extraordinary implication of the two concepts, which are literally the one within the other. The *kairos*—to translate it simply as "occasion" or "chance" would be trivial—is not another time: what we get when we grasp a *kairos* is not another time, but only a contracted and abridged *chronos*. The precious pearl in the ring of chance is only a small portion of *chronos*, a time which is left.[25]

Kairos is a temporal contraction, the time that ushers in the event. This is perhaps why Nietzsche is careful with his terminology in *BGE* 274. At first he simply refers to those who are waiting (whether they know it or not) for the right time (*zur rechten Zeit*). This is a first approximation to the problem of waiting, but when Nietzsche names the *kairos* at the end of the aphorism, the phrase "*die rechte Zeit*" appears in apposition and in quotation marks, indicating that this is at best an approach, a translation, because the time in question is not simply a passing moment, even one that has been marked for a mundane event, like an appointment in one's daily agenda. It is precisely that quickening and condensation, that unpredictable moment of turning, that cannot be scheduled. Perhaps at best we can clear away the obstacles that stand in the way of our vigilance. Much of "What Is Noble?" concerns such questions of vigilant strategy.

The philosopheme of time's contraction has appeared earlier in *BGE*. We should be alert to the emergence of such contractions; yet, Nietzsche argues, our modern virtues stand in the way of acknowledging these moments. The hypercultivation of the historical sense encourages us to place all times and events on the same plane, leaving us unprepared to "seize chance by the forelock." In "Our Virtues" he writes:

> Perhaps our great virtue of historical sense is necessarily opposed to *good* taste, at least to the very best taste, and it is only poorly and haltingly, only with effort,

> that we are able to reproduce in ourselves the brief and lesser as well as greatest serendipities [*Glücksfälle*] and transfigurations of human life as they light up every now and then: these moments and marvels [*Augenblicke und Wunder*] when a great force stands necessarily still in front of the boundless and limitless—, the enjoyment of an abundance of subtle pleasure [*Überfluss von feiner Lust*] in suddenly mastering and inscribing in stone [*Bändigung und Versteinerung*], in settling down and establishing yourself on ground that is still shaking (*BGE* 224).

Taste is understood here as a temporal sense. In the long aphorism that that these words help to conclude, Nietzsche has been discussing what he sees as a distinctively modern weakness, the European historical sense that is an effect of "the democratic mixing [*Vermengung*] of classes and races." We moderns may claim this as a sixth sense, he says. If so, the implicit point may be that it replaces taste, a more vital and crucial sense, also sometimes said to be the sixth. Nietzsche speaks of a mixing or *Vermengung* here, a becoming-multitude that is an "enchanting and crazy *half-barbarism*" leading to a weakness for the barbaric and ignoble. The century of the *Menge*, again, is the condition with which he contrasts the higher taste for the transformative moment. As so often, when Nietzsche thinks of seizing the moment, one of his leading countertexts is Hegel's world-history and his concept of the great event. Hegel sees the barbarism of the Germanic tribes as a virtue; it was their unformed and receptive nature, at the transition to the European era (variously called Christian, romantic, or Germanic) that allowed them to embrace Latin Christianity and the remnants of classical culture it brought in its wake.[26]

In a letter to Jacob Burckhardt, Nietzsche characterized *Beyond Good and Evil* as saying everything that was said in *Zarathustra*, but doing so very differently; in *Ecce Homo*, he specifies the difference, beginning the section on *BGE* by saying, "After the Yes-saying part of my task had been solved, the time had come for the No-saying, *No-doing* part . . ."(*KSB* 7.254). Reading this hermeneutic advice from the perspective of "the problem of those who wait," we could see their situation as the other side of the "great event" or "the great noon" that are promised in *Zarathustra*. Earlier I've argued that Nietzsche's great event must be understood as a great event of the earth, in contrast to the nineteenth-century's default Hegelian concept of the great event as involving the state. It is a question of the world of world-history as opposed to the human-earth of mobile habitation. Similarly, Nietzsche's futurity should be heard as the counterpart of the Hegelian "end of history." We hear one echo of these great events in *BGE* 32, where Nietzsche distinguishes the premoral,

moral, and postmoral epochs; the moral epoch first took hold with the rise of civilization ten thousand years ago, while we may now be on the verge of the postmoral period. *BGE* 285 notes how difficult it is to perceive great events, because their light reaches us only many years later, like that from distant stars. Nevertheless, Nietzsche maintains the possibility of a disciplined vigilance. It is the last human who can no longer imagine the *kairos*, and so has lost all sense of this deeper form of waiting. At best, the last men and women revere the temporary stars of the spectacle. The problem of those who wait is the problem of living in "the century of the *Menge*," in the age, *saeculum*, or tranche of *chronos* that we can also call the epoch of theater and spectacle. The distraction of the time places all of us in danger of falling into oblivion of the *kairos*. Oblivion is more than forgetting. Those who have been immersed in the river *Lethe* not only forget, but forget that they have forgotten. In contrast, *Beyond Good and Evil* enacts its *Prelude to a Philosophy of the Future*, close to what Nietzsche sometimes calls "philosophy for the day after tomorrow." It awakens the thought of serendipity, *Zufall*, *Glücksfall*, and *kairos*, while offering strategies for dissolving false identifications of, and misplaced enthusiasms for, supposed great humans and great events.

AMORTIZING TIME: THE DEBT CRISIS

I have been arguing that we should give more attention to the "old philologist" Nietzsche and his terminology. Earth and world-history, multitude and masses, *kairos* and *chronos* are pairs of terms that open philosophical questions concerning the topoi conventionally known as history, time, and the state. While most critics and commentators have ignored these, or have failed to understand the oppositional relationship in these pairs, Christ and Antichrist have, unsurprisingly, elicited much more response, even if that response is often ill-informed concerning Nietzsche's relation to Christian theology. Later I will say something about how Nietzsche, son of a ministerial family, former theology student, and housemate of the antitheological theologian Franz Overbeck, inflected these terms, especially through his understanding of political theology. First, however, I turn to the binary of debt and credit, which Nietzsche finds indispensable in uncovering the political economy of modernity, the deeper structure underlying its conventional "history."

Nietzsche described the crisis of his and our time as a crisis of debt. "I live on my own credit," he wrote in *Ecce Homo*, presumably meaning that his entire life was a promissory note that could be redeemed only by the posthumous success of his thought. In *On the Genealogy of Morality*, usually considered his most systematic book, he constructs a narrative explaining how human beings have come to be overwhelmed individually and collectively, internally and externally, by a sense of debt so great that it could be repaid only through a miracle. Nietzsche's well-known analysis of the priests (credit officers of psychic debt, themselves indebted) and of the transformations of the ascetic ideal (various ultralong-term credit plans) are developments of his political economy of debt and credit. Without retracing the entire story, let us recall some of its steps. Debt (*Schuld*) is at first a simple but pervasive social and economic fact. This primordial debt is already inscribed in all forms of exchange, and in all social bonds. The human, the *Mensch* or man, Nietzsche says, is the one who measures, drawing on an etymology from the Sanskrit *manas* (I am told by scholars that this derivation is not impossible). This measuring is a measuring of debt. Once measured, a debt might be repaid in various ways; if no money or acceptable objects are available, ancient law allowed that the creditor might torture or mutilate the debtor to gain what was deemed to be an equivalent amount of cruel joy. As a result of the amazing and catastrophic change by which humans came to live in relative peace, they adopted a morality that restricted their instincts, forcing them into an internalized life in which they were responsible to a higher power for transgressions and even temptations to transgression. This was the first "great event" in what Nietzsche calls "major history" (*Hauptgeschichte*). The newly civilized measured their own guilt and indebtedness, making bumbling efforts to repay the constantly increasing debt that is the price of civilization, finally concocting the mad religion of Christianity in which the holy one sacrifices himself to *redeem* (*erlösen*) the debt, another term that migrates from elemental economics to spiritual religion. Yet this repayment comes to be seen as generating a new debt to the redeemer, and so we spiral into greater and irredeemable indebtedness and guilt.

Debt and debt management shape political time. While much of Nietzsche's argument in the *Genealogy* is well known, exploring it with an emphasis on the question of debt or guilt (*Schuld*) demonstrates how they contribute to Nietzsche's earth-oriented challenge to "so-called world-history." His understanding of history as a process of the production, accumulation, and accounting of debt is a theory of how the debt machine constructs a future restricted to training and disciplining humans to make restitution. It is a theory of time, ultimately of theologico-political time, and one, he claims, that has infected the philosophical tradition from its beginnings. The Church comes to see *Schuld* as irredeemable without Christ's intervention. Yet the redeemer (*Erlöser*) finally serves only to remind us of the enormity of the debt, and those who have not glimpsed the Nietzschean *Unschuld des Werdens*, or "unmortgaged becoming" (usually translated as the "innocence of becoming") will die still "owing their soul to the company store."

For Nietzsche, the human itself is understood as debtor and creditor. It is surprising that Nietzsche's philosophical etymology of "*Mensch*" ("human")—almost Heideggerian in its daring, however opposed to Heidegger in its content—has generally escaped notice. Appealing to Sanskrit rather than Greek to uncover an original meaning, he extends the critique of measure with which Zarathustra began his first public talk:

> Fixing prices, setting values, working out equivalents, exchanging—this preoccupied the first human thoughts to such a degree that in a certain sense it constitutes thought: the most primitive kind of cunning was bred here, as was also, presumably, the first appearance of human pride, the human sense of superiority over other animals. Perhaps our word "*Mensch*" (*manas*) expresses something of this first sensation of self-confidence: the *Mensch* designates itself as the being who measures values, who values and measures, as the "calculating animal as such" (*GM* II 8).[27]

In contrast to a gift economy of "sacrifice" and "squandering" for the sake of the earth, Zarathustra condemns the last humans for having made everything small by their calculations of work, leisure, pleasure, and rest (*Z* Prologue 3–5). All of *Zarathustra*, as I've argued elsewhere, is permeated by an economy of the gift.[28] In the second essay of *GM*, he asserts that "the feeling of guilt, of personal obligation . . . originated . . . in the oldest and most primitive personal relationship there is, in the relationship of buyer and seller, of creditor and debtor; here person met person for the first time, and *measured* himself person against person" (*GM* II.8). Nietzsche, it should be clear, is not reading liberal economics back into "human nature" (as if he had an innocently unproblematic concept of that notion). As the treatment of debt unfolds, Nietzsche reconstructs transitions, some continuous and some more abrupt, from one-on-one debt to debts owed the community, to ancestors, eventually to debts the community owes the gods, and finally to the single god of "monotonotheism." Even more worthy of thought is Nietzsche's further claim that this relation *constitutes* thought.

The philological identification of the *human* as esteemer is one already made in *Zarathustra* in the chapter "On the Thousand and One Goals":

> Verily, humans gave themselves all their good and evil. Verily, they did not take it, they did not find it, nor did it come to them as a voice from heaven. Only

humans placed values (*Werthe*) in things to preserve themselves—he alone created a meaning for things, a human meaning. Therefore he calls himself "*Mensch*" which means: the esteemer (*Schätzende*) (*Z* I:15).

This would be the supreme instance of what Nietzsche early in the *Genealogy* calls "the lordly right of giving names" (*GM* I:2). For humans name themselves precisely as the one who gives such names insofar as they esteem and create (*Schätzen ist Schaffen*). The attempt to bolster this interpretation of *Mensch* (related to the English "man" but distinct in German from the gendered *Mann*) by reference to the Sanskrit *manas* seems no better or worse by conventional philological criteria than other such etymologies in Nietzsche and Heidegger. Esteeming and disdaining are listed in Sanskrit dictionaries as among the senses or cognates of *manas* but not as the word's primary meaning.

We know what Heidegger would say about Nietzsche's identification of man as the valuer or esteemer. He would see Nietzsche as simply projecting back into the origins of thought the value-thinking that is typical of the metaphysical era in its completion; for Heidegger, that is also the destined culmination of that era, already implicit in the metaphysics of presence. It is just such projection, he would maintain, that prevents Nietzsche from entering into thoughtful conversation with the early Greeks. And he would add that it is just this adoption of valuational thinking as the norm that also prevents Nietzsche from thinking beyond the tradition—and in fact locks him into it.[29] But is it so clear that Nietzsche has trapped himself in that way?

To say with Heidegger that Nietzsche sees no other possibility than valuative thinking, and sees man as *nothing but* the evaluator, would be to ignore the very important point that the human is a *limited* concept for Nietzsche. One might say, in the spirit of Foucault's bon mot, that the face of the human may be washed away like a figure drawn in sand, that he is the first thinker to attempt to expose and explore the limits of the concept of the human, and that the texts just cited are contributions to discerning those limits. "The human" Zarathustra announces in his very first speeches "is something that must be overcome." And if, as Nietzsche suggests, we are to understand the human as "the evaluator," then how ought we to understand and translate "*Übermensch*"? As "metaevaluator," as "human beyond evaluation," as "posthuman" or as "postevaluator"?

In *The Birth of Tragedy* Nietzsche had spoken of the truth as that which lies beyond measuring, when he said of the Dionysian that "excess (*Übermass*) reveals itself as truth" (*BT* 4). The *Übermensch* is the excessive one who goes beyond measure, which means that he goes beyond the human as the measure or measurer. This way of understanding the *Übermensch* may, incidentally, help to distinguish Nietzschean thinking from the sophistic or Protagorean relativism with which it is often all too hastily associated. To the extent that the human is not ultimate, the force of the familiar "man is the measure of all things" is drastically undercut (I give the usual translation; Protagoras had, of course, spoken of the less-gendered *anthropos*). Nietzsche is not only reaching back to prehistory (*Vorzeit*), as he tells us, but also to the archaic origins of philosophy. According to tradition, the first recorded sentence of Greek philosophy is Anaximander's saying, which Nietzsche had translated in 1873 as:

> Whence things have their origin, there they must also pass away according to necessity; for they must pay penalty and be judged for their injustice, according to the ordinance of time. (*Woher die Dinge ihre Entstehung haben, dahin müssen sie auch zu Grunde gehen, nach der Notwendigkeit; den sie müssen Büsse zahlen und fur ihre Ungerechtigkeiten gerichtet warden, gemäss der Ordnung der Zeit* (*PTG* 45; *KSA* 1.818).

As various commentators have observed, Anaximander's language may very well owe a great deal to the commercial society of the eastern Mediterranean, in which the city of Miletus was an active hub.

Let us read three significant passages where Nietzsche confronts Anaximander directly or by allusion. In each passage Nietzsche emphasizes and articulates the themes of debt, penalty, time, and punishment that Heidegger wants to eliminate through his own reading of this saying. The most comprehensive treatment of the economy of thought, practice, and culture in these terms is that in the *Genealogy*. But the *Spruch* or saying itself is not in need of rescue, for the first thing that Nietzsche says about Anaximander in *Philosophy in the Tragic Age of the Greeks* is that his sentences are quite in order as they are. They are not mere "fragments." He is described as "the first philosophical author of the ancients," who "writes exactly as one expects a typical philosopher to write when alienating demands have not yet robbed him of his innocence and naiveté. That is to say, in graven stylized letters, sentence after sentence the witness to fresh illumination, each the expression of time spent in sublime mediation" (*PTG* 45).

This is a remarkable claim to make on the basis of one surviving sentence. From this perspective, the text we are dealing with is not a fragment in need of rescue, but a sentence hanging over our heads. To Anaximander's image as the first to speak of the *arché* we can now add that he is the first to put philosophy into writing, and to inscribe it with a force such that later thinkers will necessarily be indebted to it. This lapidary inscription forever marks the body of philosophy. Like the marks of punishment Nietzsche describes in the *Genealogy*, they provide a forced, perhaps painful, memory that seems inescapable. After quoting the sentence—a sentence that Nietzsche says speaks with "lapidary impressiveness"—he wonders how we are to read it: "Enigmatic proclamation of a true pessimist, oracular inscription [*Orakelaufschrift*] over the boundary stone of Greek philosophy: how shall we interpret you?"(*PTG* 45–46).

At this point Nietzsche quotes Schopenhauer, who as "the only serious moralist of our century charges us with a similar reflection":

> The proper measure with which to judge any and all human beings is that they are really creatures who should not exist at all and who are doing penance [*Büsse zahlen*, the same words that appear in Nietzsche's translation] for their lives by their manifold sufferings and their death. What could we expect of such creatures? Are we not all sinners under sentence of death? We do penance for having been born, first by living and then by dying (*PTG* 46).

Nietzsche thus constructs a metahistory of philosophy in his meditations on the early Greeks, a metahistory based on debt to be repaid in time, and so making life a time of paying one's debts. It should be contrasted with the Heideggerian metahistory (beginning with Plato) of philosophy as the metaphysics of presence. Insofar as the debt always remains to be paid, the philosophers in Nietzsche's metahistory always find presence elusive.

A second *Auseinandersetzung* with Anaximander occurs in the *Zarathustra* chapter entitled "*Von der Erlösung*" or "Of Redemption" (*Z* II:20). Let us note again that *Erlösung* is a multifaceted word that can designate religious or spiritual redemption on the one hand or the redeeming of a debt on the other. And in this chapter, redemption is considered on several levels. The issue is introduced by a hunchback spokesman for a number of "cripples," who says that Zarathustra cannot persuade the people unless he also persuades the blind and deformed. If he could correct or redeem their bodily excesses and defects, he would be a more plausible teacher. As noted earlier, Zarathustra' interlocutor offers him a *kairos*, an opportune moment to achieve greater respect and authority. Of course this recalls the miraculous cures attributed to Jesus and other healer-magicians of his era. But, Zarathustra asks, would this in fact be a great redemption? Even the people say that to take away the hump from the hunchback is to destroy his spirit.

This thought leads to a consideration of bodily and spiritual fragmentation that now appears to be almost universal, extending far beyond obvious cases of deformity. Some people are nothing but particular bodily organs in a monstrously enlarged state, with a corresponding atrophy of others; such an inverse cripple might, for example, be an "ear as big as a man," with "a tiny envious face." So most people are nothing but fragments and severed limbs (*Bruchstücken und Gliedmassen*). Those who appear to be whole are actually fragments, while ancient sayings that appear to be fragmentary, like Anaximander's, are in fact lapidary utterances whose inscriptions remain hanging over us for millennia.

In fact the inscriptions themselves may help to account for the human fragmentation. For the essential question of redemption has to do not with the blind and the lame but with the woundings, scarrings, and divisions effected by time—and especially by the time of the inscription of revenge, a time that Nietzsche here calls madness (*Wahnsinn*). Because the will cannot will backward, because it is bound to the law of time and time's "it was," life is a perpetual process of fragmentation in which the past seems to be nothing but a collection of dispersed and shattered ruins. In this situation the will can be nothing but an "angry spectator" who sees "man in ruins and scattered as over a battlefield or a butcherfield" (*zertrümmert und zerstreuet wie über ein Schlacht- und Schlächterfeld*) (*Z* II:20).

But these fragments can also be seen as "fragments of the future"—if redemption is possible. Already Nietzsche appears to speak the language of Anaximander: the fragments (humans or human parts in this case) suffer by being cut off from the whole, perhaps simply for coming into separate existence "according to the ordinance of time." The doxographers, even if unreliable, spoke of the possibility of a redemption through the collapse of the individuated things and elements back into the whole.[30] Zarathustra now defines what he sees as the only possible form of redemption: "To redeem those who lived in the past and to recreate all 'it was' into a 'thus I willed it'—that alone should I call redemption" (*Z* II:20).

True redemption would be redemption from the spirit of revenge. Zarathustra at this point distinguishes "revenge" and "the spirit of revenge" in a way that will help in explaining the significance of Anaximander's saying: "Verily, a great folly [*Narrheit*] dwells in our will; and all men are under a curse insofar as this folly has acquired spirit" (*Z* II:20). The folly of revenge is one thing, but it becomes malevolent and dangerous when it acquires spirit. How did revenge acquire spirit? Zarathustra's answer is that thought and reflection have been preoccupied with suffering and punishment, and that "madness" (*Wahnsinn*, a step beyond *Narrheit*) has produced a law, a formula, and an inscription that has marked the will. Now this fateful inscription, I want to suggest, is the lapidary utterance of Anaximander as it has been carved and engraved in the stones and monuments of philosophy: when madness comes to preach and inscribe, then revenge passes from its simple state to one in which it has acquired spirit. Revenge masks itself as the demand for repayment of debt.

We need to read Zarathustra's comments on the spirit of revenge (at some length) so that we hear in them the resonance of the thinker and speaker of the *arché*. For the *arché* here is the principle of philosophy, the tradition from Anaximander to Schopenhauer:

> The spirit of revenge, my friends, has so far been the subject of man's best reflection; and where there was suffering, one always wanted punishment too.
>
> For "punishment" is what revenge calls itself; with a hypocritical lie it creates a good conscience for itself.
>
> Because there is suffering in those who will, inasmuch as they cannot will backwards, willing itself and all life were supposed to be—a punishment. And now cloud upon cloud rolled over the spirit, until eventually madness preached, "Everything passes away; therefore everything deserves to pass away. And this too is justice, this law of time that it must devour its children." Thus preached madness.
>
> "Things are ordered morally according to justice and punishment. Alas, where is redemption from the flux of things and from the punishment called existence?" Thus preached madness. (*Z* II:20)

The third confrontation with Anaximander is the text in the *Genealogy* where Nietzsche returns to the double theme of indebtedness and redemption. It is perhaps with reference to the "teaching of madness" that Nietzsche writes there of the "*redeeming* man of great love and contempt," a redeemer who will lift the "curse of the hitherto reigning ideal" (*GM* II:24). Here at the end of the second part of the *Genealogy* Nietzsche speaks repeatedly of redemption and of a redeemer—just after he has attempted to demonstrate that the Christian notion of redemption, tied as it is to the economic complex of debt and credit, makes the earth into a madhouse.

Among the principal themes of the *Genealogy* are guilt, debt, punishment, justice, and redemption. Nietzsche's treatment of these themes is usually read as an account of those sociopolitical formations that eventually produce the aberrations of Christianity and other forms of asceticism.[31] Yet the analysis goes further, for Nietzsche claims that such notions are so rooted in human beings that they constitute "thinking *as such*" (*das Denken*): "Setting prices, determining values, contriving equivalences, exchanging—these preoccupied the earliest thinking of man to so great an extent that in a certain sense they constitute thinking *as such*" (*GM* II:8; Nietzsche's emphasis).

The *Genealogy* is, among other things, a text about interpretation. We are told in the preface, for example, that the entire Third Essay is an *Auslegung* of a single aphorism (apparently its opening one). We could think of the Second Essay, " 'Guilt,' 'Bad Conscience,' and the Like," as an interpretation of the saying of Anaximander. Just as in Heidegger's essay, there is a sequential treatment of the significant words or concepts of the saying, and an attempt to trace the way in which the tradition inscribes its translations of these words and of the saying itself. In the course of his analysis Nietzsche, too, proposes a number of translations or equivalences—for example that between *Schuld* (guilt) and *Schulden* (debts). A similar relation obtains between the English "owe" and "ought." The principle here, Nietzsche says, is "the idea that every injury has its *equivalent* and can actually be paid back, even if only through the pain of the culprit" (*GM* II:4). He gives this observation an Oedipal, genealogical, and tragic twist by noting that this is the way that "parents still punish their children."

The question is whether the sort of thinking embodied in Anaximander's saying is inside or outside the metaphysical tradition. In his efforts to place it outside, Heidegger is forced to eliminate or retranslate a whole set of terms—*chreon, tisis, dike, taxis*—that have to do with what we may generally call the economic: the world of valuing and evaluation. For Nietzsche the position represented by Anaximander is the earliest form of civilization to which we have access, and is at least as old as the concept of "legal subjects" (*Rechtspersonen*) (*GM* II:4). Eventually, of course, we will have to ask whether there is anything more archaic than this "prehistory." We could see the claim about "thinking *as such*" as an extension of Nietzsche-Zarathustra's view of what madness preaches. As in the latter account, there is a distinction to be made between an unreflective thinking in terms of guilt, punishment, debt, and credit and the appearance of specific ideals and doctrines—codified as philosophy—based on that thinking.

According to the narrative given in the *Genealogy*, more or less reflective moralities result from a cataclysmic event brought on by the needs of organized social life and the self-inflicted transformations that warriors unwittingly incur when they bind themselves to a world of law. This is a crisis of internalization, in which guilt, debt, and punishment are no longer inscribed merely on human bodies but in human consciousness. When restricted to their consciousness, their instincts of aggression turn inward and generate an internal economy of debt and credit. Part of such an internal economy is the development of explicit religious and philosophical teachings—that is, the internal inscription of what madness preaches from Anaximander to Schopenhauer. The internalization of guilt and debt is followed by its infinitization, when the community comes to seem all powerful and is metaphorically represented by an infinite and omnipotent god. Nietzsche could be read as making explicit the economic principle that is only implicit in the sophist Protagoras. His own concepts of gift-giving, generosity, and hospitality could be developed to demonstrate that he has an alternative approach to the economic, one that could be deciphered in the hieroglyph of "*das Unschild des Werdens*" or "unmortgaged becoming." For now, however, I want to focus on how Nietzsche gives another alternative analysis, one that rewrites "so-called world history." Hegel takes world-history to be the development, articulation, and unfolding of spirit; for him it is a dialectical process and a story of Christian *parousia*. Nietzsche gives a rather abbreviated analysis of a parallel sequence of events—prehistory, early civilization, Greece and Rome, European Christianity—that tells not of the self-inflation of the spirit but of how political and religious structures lead to an impossible debt, where individuals and states all die without amortizing their underwater mortgages. Insofar as they exist in time they are always struggling to pay their debts.[32]

To live life in debt is to measure out time in terms of the conditions of repayment or the penalties of nonpayment ("time is money"). This week, month, or year, I must meet the demands of the bank, the credit card company, or the loan shark. Perhaps I balance my finances enough to support my "lifestyle." Still, each month brings new expenses that must be paid. As long as I'm doing this anyway, I might as well live a little larger, and put a few more things on credit. I aspire to a good credit rating. Now I am indebted to myself, in a sense; I must work to produce money for the month's bills. Or (in the typical life sequence touted by our economic managers), I manage to invest wisely enough to retire from employment after sixty or seventy years of preparation for navigating the world of debt and credit. Now I can do what I've always wanted, I'm told. I've escaped from the rat race, wage slavery, or debtors' prison. Yet what I am, and what I tell myself I've always wanted, has been formed by the sites that I imagine I've escaped. And of course I will probably be too old, my health too problematic, to fully enjoy these things. The indebted life is a prescription for experiencing time as *chronos*. This is not only my individual problem. Nietzsche read and apparently alludes to Emerson's deceptively brief and simple essay "Gifts." This metaeconomic theory also seems to reference Anaximander and his tradition, which could stand as an emblem of world economic crisis: "It is said that the world is in a state of bankruptcy; that the world owes the world more than the world can pay, and ought to go into chancery and be sold." Debt is the world's chronic problem, testified to by such current phrases as "sovereign debt" and the latest economic news. Might a different economy be fostered on the earth? Nietzsche promised an *Umwertung aller Werte* that might answer this question. The promise was not fulfilled (or was condensed into the single book *The Antichrist*). Yet he did manage to raise the question.

As Deleuze and Guattari observe, Nietzsche's analysis of debt in the *Genealogy* offers an important correction and contrast to traditional economic theories. If liberal and neoliberal theorists take exchange between autonomous agents to be the primary economic relationship, and Marxists give a parallel priority to social production, Nietzsche claims that the first personal relationship is that of debtor and creditor. The first model has been accurately criticized for being hopelessly ahistorical, insofar as it fails to see first that it bases its model on a historical event, the dissolution of feudal relationships. More seriously, it imposes a problematic grid on the agents of exchange, ignoring specific relations of power and culture. Most importantly for our analysis, it sees the exchange as completed and fulfilled in the moment of its execution. The liberal model presupposes a conception of time as a series of instants that can be marked by dated transactions.

Nietzsche continues his analysis of the debt economy by charting a series of social and historical variations and expansions of the debtor/creditor relationship. The individual becomes the debtor of the community; those who refuse to recognize their communal debt are subject to punishment or banishment (*GM* II.9). In a methodological interlude before again taking up his narrative of how debt and credit structure both thought and practice, Nietzsche discusses the historically polymorphous character of punishment, which is, among other things, the way in which the community collects its debt (or the equivalent) from the deadbeats who defy it. (He notes that the institution of punishment also allows the community to demonstrate its mercy, by occasionally choosing to forgo punishment.) Here Nietzsche formulates in lapidary style some of his crucial genealogical principles, which can be contrasted with the models of evolutionary or progressive development characteristic of Hegel and his intellectual progeny like Strauss and Hartmann:

> Every purpose and use is just a *sign* that the will to power has achieved mastery over something less powerful . . . the whole history of a "thing," an organ, a tradition, can to this extent be a continuous chain of signs, continually revealing new interpretations and adaptations . . . The "development" of a thing, a tradition, an organ is therefore certainly not its *progressus* toward a goal, still less is it

> a logical *progressus* . . . The form is fluid, the "meaning" even more so . . . only something which has no history can be defined (*GM* II.12, 13)

Genealogy differs from "Hegelian" history in being neither teleological nor logical. This methodological clarification comes in the midst of Nietzsche's tracing the transformations of the credit/debt structure, which yields an alternative account of such things as the emergence of the state and of the universal religion of Christianity, precisely the crucial factors in "so-called world-history."

How did the state arise? Nietzsche's well-known account summons up a scenario of an aggressive, warlike, nomadic band falling upon and mastering a weaker, passive, sedentary human group (*GM* II.16–17). It is not a principle of cultural or spiritual unity, as the idealists would have it, nor is it a contract, but the unintended consequence of a rather arbitrary collision. Once the conquest has been achieved, it can be maintained only through organization and discipline, a discipline requiring that the former marauders rein in their aggressive instincts under some authority (the state). Instincts, however, are not easily extirpated, so "all those instincts of the wild, free, roving human were turned backwards, *against the human itself.*" The "involuntary artists" who generate archaic political states must redirect their aggression, and they do so against themselves, in a process by which one "gives form to oneself as a piece of difficult, resisting, suffering matter, branding it with a will, a critique, a contradiction, a contempt, a 'no,' this uncanny but joyous labor of a soul voluntarily split within itself . . ." (*GM* II.18). This is the invention of bad conscience (*schlechtes Gewissen*) and the transformation of *Schuld* from a more or less material *debt* to another person or group to psychological *guilt*, an internal sense of having gone wrong, of having sinned simply through one's desires, whether acted on or not.

With the emergence of a powerful community enforcing law and order, there comes a second transformation of the debtor/creditor relationship. What was primitively a relation among individuals, and then one between the individual and the community, now takes the shape of a relation between the community and its founders or guarantors, first thought of as ancestors, but eventually, with the growth of state power, elevated to the status of gods (*GM* II.19). In this sequence of social forms, Nietzsche betrays his own preference for what he calls the "middle period" of "noble tribes." But from here the state consolidates its identity even further by abandoning or subordinating the principle of "blood lineage," which it had inherited from its tribal beginnings, constituting itself as an association of subjects or citizens (this corresponds to what Deleuze and Guattari call the supersession of coded, segmented societies by the overcoding of despotism[33]). Hegel, we recall, firmly distinguishes mere *Nationen*, linked only by a common network of descent, from true states. But if Hegel sees the distinguishing factor in a spiritual principle (in his sense), Nietzsche finds it in the necessity for the more unified community to deal with the presence of those who are marginal yet indispensable, "slaves and serfs," for example, who seem to straddle the inside/outside divide.[34]

Nietzsche's analysis of *Schuld*, its political economy, and its ideological apparatuses, amplifies Zarathustra's railing against the last humans with their calculated, secured, and insured routines of pleasure, work, and self-maintenance. For Nietzsche—who sometimes represents himself as a physician—the excessively measured world of debt and credit, monetary or idealized, is a *chronic* disease in a double sense. It both recurs over a long stretch of time and is also itself a disorder or malaise of temporality. "Time is money"—yes, Nietzsche would say, so it is *now*, in European modernity (or capitalism). Yet money itself is simply one of the forms that the debt/credit economy takes in the *Hauptgeschichte* of the calculative *Mensch*. Debt precedes money, even if money has become its overwhelmingly pervasive mode. The question posed by this history is not only how you can free yourself from debt, but from debt's regime of temporality. This would be the true redemption or *Erlösung*. Unfortunately, Nietzsche did not himself engage in the dialogue with Marx on time, history, and temporality that the subject calls for. Yet the dialogue has been conducted (if not always labeled as such) by Weber, Foucault, Deleuze, Sloterdijk, and others. As we will see in the following chapters, in the place where Nietzsche might have begun to think *kairos* and *chronos* in relation to concepts such as commodification, the falling rate of profit, or economic imperialism, he turns toward two other approaches to the earth. At first these two seem rather disjunct. On the one hand, Nietzsche imagines the earth as a garden, a space of cultivation, abundance, and beauty. On the other, he attempts a final undermining of the mutually dependent beliefs and practices of Christianity and world-history, seeking to expel both for the sake of the human-earth by his complexly parodic identification with the figure of the Antichrist.

CHAPTER 5

"The World Awaits You as a Garden": A Political Aesthetic of the Anthropocene?

And the Lord God planted a garden eastward in Eden; and there he put the man that he had formed.

GENESIS

In painting, in sculpture, indeed in all the visual arts including architecture and horticulture insofar as they are fine arts, design is what is essential.

KANT, *Critique of Judgment*

Every living thing needs to be surrounded by an atmosphere, a mysterious circle of mist: if one robs it of this veil, if one condemns a religion, an art, a genius to orbit as a star without an atmosphere: then one should not wonder about its becoming withered, hard, and barren.

NIETZSCHE, *On the Advantage and Disadvantage of History for Life*, 7

If *Thus Spoke Zarathustra* is Nietzsche's song of the earth, his philosophical landscape poem, shouldn't we expect it to indicate or foreshadow in some way earth's great event, or "that stroke of midday and of great decision that makes the will free again, which gives earth its purpose and humans their hope again"? (*GM* II.24). In this chapter I ask whether Zarathustra's animals adumbrate an answer. The earthly snake and the aerial eagle, singing in childlike chorus, tell him in their charming, chattering way that for Zarathustra at least—and for the future human-earth insofar as he is its forerunner (*Vorläufer*)—"the world awaits you as a garden." The garden, in Western tradition, has also promised an escape or respite from worldly time—a site of refuge, contemplation, amorous activity, a place for dolce far niente. It offers an earthly happiness somewhat removed from the world, a form of what Foucault calls a heterotopia. Moreover, Zarathustra had earlier described his "children" as trees in a plantation, so the garden awaiting him can also be seen as his legacy to the future (*Z* III.3).

NIETZSCHE'S HUMAN-EARTH: LESS THAN NATURE, MORE THAN WORLD

What might the earth become? What is the earth to which Nietzsche's Zarathustra enjoins us to be true? How can the earth become the site of transformative events? In preceding chapters, I have argued that Nietzsche's "earth" is narrower than "nature" and distinct from what many of his contemporaries and predecessors mean by "world." Earth is not all of what there is but the immanent site of human existence. To speak of the earthly is certainly to speak of embodied action and passion, known and experienced, as opposed to an imagined otherworld. More specifically, I have been exploring the aspect of what Nietzsche calls the "human-earth (*Menschen-Erde*)," the aspect under which it is the site of mobile human habitation, rather than the "world" understood as a complex of states within the perspective of world-history. The future and futurity of this earth, as we have seen, is the leading focus of Nietzsche's political hopes. Yet Nietzsche's readers often feel that he provides only a few tantalizing clues to the nature or characteristics of this earth's future, beyond its contrasts with the self-image of the "world." We can say that earth will be a home for the enigmatic figure of the *Übermensch*, a figure defined by its own radical futurity, a futurity set free in the innocence of a becoming without debt (*Unschuld des Werdens*). Is there something more to be known about the form of the earth's futurity? How could the earth be transformed so as to become the ground or theater of great events?

One significant thing we do learn about this earth is that it presents itself as, or becomes, a garden. An intense expression of this dimension appears in Zarathustra's transvaluation of sensuality (*Wollust*) in "The Three Evils": "Sensuality: for free hearts innocent and free, the garden-happiness of the earth, all futures' exuberance of thanks to the now." The sensual happiness associated with the garden motif is traditional. In a more distinctively Nietzschean vein, this happiness is very explicitly said to be not merely a present foretaste of future joy but a retroactive effect of the future upon the present. It is as if the future is drawing forth its own preparation. This temporal paradox is pervasive in Zarathustra, although it has been little explored by commentators.[1]

Even more to the point, when Zarathustra emerges from his convalescent swoon, he explicitly accepts *only* that part of his animals' chatter in which they declare that "the world awaits you as a garden" (the animals speak of the *Welt* rather than the *Erde*; but they are chattering or babbling—*schwätzen*—as Zarathustra says). And in *Beyond Good and Evil*, where Nietzsche offers provisional pointers to free spirits, budding philosophers of the future, and good Europeans who are vigilantly guarding their spiritual resources and watching for opportune moments in the century of the multitude, he counsels "please don't forget the garden, the garden with the golden trellises!" (*BGE* 25). If Nietzsche's late persona, the Antichrist, is drawn from the conclusion of the Christian narrative schema that was filled in by world-history, the topos of the garden has a Biblical resonance from that story's beginning, an iconic image of the "innocence of becoming." Human beings are given their natural place in the Garden of Eden, where they have dominion over the earth and its animals. We might say that Nietzsche's figure of the garden rewrites this narrative and its topos on an earth without God or sin. In this vein, Nietzsche's correlation of garden-happiness with sensuality also rewrites the usual understanding of the Biblical garden story as one in which the fall of the primal pair involves their discovery of their nakedness and shame over their own sensuality.

At the same time, we can understand the garden in terms of an updated interpretation of Nietzsche's *Menschen-Erde*. Consider the human-earth as what the earth has become in the geological era now called the anthropocene, in which humans continue to alter the earth through activities such as rapid population increase; domestication of animals; industrial agriculture and fishing; mining and drilling; large-scale mineralization of the surface; burning of fossil fuels; chemical change of the atmosphere, soil, and oceans; and of course (and to a large extent consequently), global warming and climate change. While Nietzsche was not aware of many implications of these dimensions of the anthropocene (any more than were his typical contemporaries), his talk of the human-earth assumes an earth that has been and continues to be transformed by human habitation and industry. There is no pure or natural earth to look to as a norm; Nietzsche knows enough of geology and evolutionary theory to avoid being beguiled by that fantasy. From this perspective, his idea of earth as a garden (or gardens) is an attempt to sketch an aesthetic politics of the human-earth or a geoaesthetics and geopolitics of the anthropocene. I take my point of departure here from *Thus Spoke Zarathustra* and its rich garden imagery, using this topos to draw on related appearances of gardens and gardening in additional Nietzschean texts.

EARTH'S GARDEN-HAPPINESS

In *Thus Spoke Zarathustra* Nietzsche's eponymous hero begins his teaching by calling on anyone who will listen—a miscellaneous audience in the marketplace—to be loyal and faithful to the earth. He urges them to turn away from all fantasies of an extra-earthly world and to reject the all-too-limited and measured satisfactions of the last humans. Yet what shall the character of the earth be when freed from such distortions, and cultivated with care and passion? Nietzsche's most consistent name for this transformed earth is "garden," and all of *Zarathustra* can be understood as an extended landscape or garden poem; it is surprising how little attention has been given to this dimension of what the author considered the "greatest gift" ever given to humanity.[2] The theme becomes explicit in the conversation of Zarathustra and his animals as he is aroused from a seven-day coma, after struggling with his "most abysmal thought." This chapter on "The Convalescent" is typically read as one of Nietzsche's most articulate and comprehensive accounts of the thought of eternal recurrence. Here his proud eagle and cunning snake join in a series of speeches describing the Great Year of Becoming that they understand to be Zarathustra's teaching. While many readers take the animals' words to be a definitive statement of that teaching, several note that Zarathustra calls them "buffoons and barrel-organs," chiding them for turning his thought into a "hurdy-gurdy song" (*Leierlied*); he also compares the animals' "chattering" (*schwätzen*) to music, which bridges the otherwise insurmountable distance between souls. Some readers attempt to reconstruct the thought by freeing it from the animals' misunderstandings of concept and tone. Here I refrain from this project of reconstruction, which many (including myself) have pursued in other contexts. I focus rather on the one thing that Zarathustra clearly accepts without reservation in what the animals tell him in their sing-song way: "the world awaits you as a garden." Here Nietzsche's words (more specifically, the animals') apparently deviate from the distinction between world and earth that I have proposed. I acknowledge that the terminology is not completely consistent, but I think it is clear here, in the solitude of Zarathustra's mountain retreat, that he and the animals are not speaking of the metaphysical world of states (as in Hegel), and the animals' whole tone is reminiscent of children's rhyme or song.

The declaration that the garden awaits is made twice and gladly received. While "The Convalescent" chapter of *Zarathustra* is the subject of frequent commentary, because of its relevance for the thought of eternal recurrence, very few discussions attempt to take stock of the garden theme announced and embodied here. This is puzzling, not only because Zarathustra twice affirms the idea that the world now awaits him as a garden, but also because the narrative *describes* a vivid garden-like scene. While Zarathustra lies silent, convalescing for seven days, his eagle has assembled a variety of plants and animals for him, creating a lush garden atmosphere:

> The eagle flew off to fetch nourishment. And whatever he collected from his plundering he laid on Zarathustra's pallet, so that Zarathustra eventually lay under a heap of yellow and red berries, grapes, rose-apples, fragrant herbs, and pine-cones. And at his feet two lambs were spread out, which the eagle had with difficulty stolen from their shepherds.

The multisensory atmosphere of this gardenesque scene involves a play of colors, fragrances, breezes, and invitations to the touch and taste. Zarathustra's animals provide the music, and refer to the singing of the songbirds. The parodic relation to the Biblical story of Adam awaking in the garden (paradise) created by God should be evident; here it is Zarathustra and his animals who have made this garden-happiness possible. Zarathustra's first act upon rising was to take a rose-apple in hand, smell it, and delight in its fragrance. This act frames his entire dialogue with the animals, who tell him, "Step out from your cave: the world awaits you as a garden. The wind is playing with heavy fragrances that would come to you; and all streams would like to follow you."

Now Zarathustra speaks for the first time since wrestling with his most abysmal thought, imploring the animals to chatter on, for "where there is chatter, there the world lies for me as a garden." He expresses gratitude for the animals' "words and tones," which act as "rainbows and seeming-bridges" between his world and the outside. The animals reply in chattering chorus, articulating a version of the thought of recurrence. This provokes Zarathustra to accuse them of turning it into a "hurdy-gurdy song," clearly a musical form inferior to that of their initial soothing "words and tones." He then relates the great pain and nausea he experienced in thinking through the inevitable return of the small human, which reminded him of the soothsayer's nihilistic prophecy he had received earlier. "The human-earth (*Menschen-Erde*) became for me a cave," he explains, vividly describing it as a place of graves, bones, and decay. This conversation sets up a contrast between the wasteland earth earlier prophesied by the soothsayer—which the latter had described in gruesome and graphic detail—and the earth as a garden announced by the animals and affirmatively welcomed by Zarathustra. Again the animals urge Zarathustra to "go out to where the world awaits you as a garden . . . to the roses and bees and the flocks of doves! But especially to the songbirds" so that he may convalesce by learning to sing—and sing he does in the following chapters—songs of the earth. Yet once again the animals attempt to shift the topic to the thought of recurrence, while Zarathustra remains focused on the meaning and experience of the earth. He enjoys the apple in all innocence, in contrast to the sin recorded in Genesis.[3] Now they produce an even longer account of what they claim to know that he teaches, a beautiful account of his supposed teaching of a Great Year of recurrence. This prosopopoeia reaches a climax when they imagine Zarathustra proclaiming "the word of the great earth and humans' noon (*grosse Erde- und Menschen-Mittags*)" before he announces his own death as a herald. So two visions of the earth are juxtaposed in this chapter, an earth of death, decay, and silence as opposed to one rich in vegetation, animal life, birdsong, attractive fragrances, and tempting food. If Zarathustra is to die, it will be as a herald in the garden rather than as a depressive prisoner in a gloomy cave or wasteland. Zarathustra twice affirms the animals' announcement that earth awaits him as a garden, but pointedly declines any engagement with their repeated versions of eternal recurrence and of what he supposedly must be thinking and what he supposedly would say. While I do not discount the importance of the animals' speech and chatter (yet what *does* it mean when an animal speaks?) and do not ignore the importance of this crucial chapter for understanding Nietzsche's idea of recurrence, I want to underline the fact, passed over by many readers, that the appearance of this thought is framed by and entangled in the confrontation between the soothsayer's vision of a deathly, standstill earth and the bountiful and welcoming garden actually seen by Zarathustra and his animals. In the three following chapters, which conclude Part III of the work (and plausibly, in a narrative perspective, the work as a whole[4]), Zarathustra follows the imperative to sing, as he serenades his soul, dances with a personified Life, and finally, in "The Seven Seals," pledges his troth to Eternity. Schematically, these songs move from the internality of the soul, to a dialogue with another, to the cosmic geoaesthetics of decentered land, sea, and sky that reveals itself to Zarathustra's "bird wisdom."[5] Zarathustra has learned from the song-birds, finally achieving a virtual avian metamorphosis.

It is surprising that there has been minimal recognition of the emergence of the garden theme at this crucial point in Nietzsche's story. If "The Convalescent" does have an exemplary significance for understanding both the book in which it occurs, and so for Nietzsche's thought, as all commentators apparently agree, then these questions arise: What is the meaning of the garden? How does the concept (or metaphor, if you prefer) inflect our reading of this signature text and others? Can the garden motif contribute to understanding what Nietzsche's Zarathustra calls the direction of the earth?

THE GARDEN AS ART AND THE GENEALOGY OF AESTHETICS

We can begin to approach these questions by reviewing some of the traditional meanings and associations of gardens—philosophical, literary, and historical—that Nietzsche draws upon. The garden has often been a powerful symbol of the union or fusion of the natural and the cultural; yet we will see that Nietzsche is quite suspicious of what he calls "sentimental" attitudes toward this supposed fusion, which might be better understood not as unity but as assemblage or combination. As a deliberate shaping and arranging of earthly elements, it creates and establishes meaning on the *Menschen-Erde*. It is a form of thinking with the earth, a realization of its possibilities, and a mode of giving the earth a direction (*Sinn*). The garden can be seen as a descendant of the great design projects of early urbanizing cultures, whose emergence marked the "great event" of the moral phase of human history on the earth (*BGE* 32). In the Preface to *Beyond Good and Evil*, Nietzsche refers to the monumental "architecture in the grand style" of ancient Asia and Egypt, suggesting that "all great things have to bestride the earth in monstrous and frightening masks in order to inscribe themselves in the hearts of humanity with eternal demands." The demand dimly foreshadowed in archaic landscape architecture and monumental design might be that of giving the earth a direction on the largest scale, while the astrological despotism, which typically provided its ideological justification, is the monstrous mask it was forced to assume. Marking and transforming the earth, the garden invites and calls for present joy while presenting a model for the future. In *Dawn* Nietzsche exclaims with horror at another tattooing of the earth: "what a wretched place Christianity has managed to make of the earth, merely by erecting the crucifix everywhere, thereby branding the earth as the place 'where the righteous are tortured to death'!" (*D* 77).

Nietzsche's concept of the garden can be articulated in terms like those that Heidegger used to analyze the originality of the work of art, namely the way in which it serves as an *origin*. Heidegger speaks of the Greek temple, for example, as opening up a world of meanings and directions on the ground of an earth that emerges through the work's act of rendering and framing, while retaining an irreducible dimension of resistance and closure.[6] However, Heidegger's conception of the world, at least in his essay on *The Origin of the*

Work of Art, is associated with ideas of state-founding and sacrifice, and so is closer to Hegel's thought than to Nietzsche's. While Heidegger has nothing explicit to say about the garden, Hegel systematically marginalizes it. In much the same way that his *Philosophy of World History* takes geography to be simply a frame or presupposition of history, his *Lectures on Aesthetics* treats gardens only as an addendum to architecture, in his system the most material of the arts. He marginalized garden art in a small appendix to architecture, minimally recognizing its value as a frame for architecture, but adding that even if one strolls delightedly through a grand park, one would never want to visit it twice (perhaps one of his classroom witticisms). Even analytic philosophers are much more beholden to Hegelian aesthetics than they suspect, because they have generally bought into the official post-Hegelian canon, enshrined in textbooks; in his view, they are at most mere pleasant frames of buildings, and never warrant a second visit.[7]

Perhaps we can take a step back, before Hegel and Heidegger, to articulate Nietzsche's idea of the garden more fully with the enormous and largely forgotten role that the concept plays in the formation of modern aesthetics. The peculiar history or genealogy of aesthetics, which did not really take on anything like its modern form before the eighteenth century, has been investigated by a number of scholars, including Heidegger's student Paul Oskar Kristeller. In his 1950 essay "On the Modern System of the Arts," Kristeller, in what I would call a proto-Foucauldian vein, shows that aesthetics as we understand it—a unified theory of natural and artistic beauty and related attributions that are correlated with a distinctive faculty of "taste," as it was called by Hume and Kant—emerged only in the European eighteenth century.[8] It took that long before it was possible to conceive all of today's "fine arts" as belonging together and properly apprehended in a way that was neither simply cognitive nor practical. [9] In its eighteenth- century emergence, aesthetics, in England, France, and Germany, saw the landscape garden as a major form of art. Kant adopted this view in the *Critique of Judgment*, even if he classified it as a form of painting. What this eighteenth-century discourse shows is that before the rapid ascendancy of the museum in the next century, the garden was the most prominent large-scale space for aesthetic viewing. Treatises on garden art and controversies between advocates of the French style and the more "natural" English one, and further polemics among various versions of the picturesque, gave rise to lively discussions. The critic Horace Walpole spoke for the age in saying that "Poetry, Painting & Gardening, or the science of Landscape, will forever by men of Taste be deemed Three Sisters, or *the Three New Graces* who dress and adorn nature."[10] The sense of the earth was a subject of aesthetic contemplation in the garden. The practice of garden design and construction was respected on a level similar to that accorded poets and painters.[11]

Yet as Elizabeth Barlow Rogers notes in her magisterial book *Landscape Design: A Cultural and Architectural History*, by 1830 thinking about gardens was displaced from philosophy to garden manuals.[12] These now tended to focus more on botany and horticulture than on large-scale design that aimed at creating a meaningful theater of the elements. As Rogers observes, the change was accompanied both by a feminization of gardening (which then entailed relegating it to the status of an allegedly minor or quasi-art) and to the practices of a newly affluent and leisured middle class (rather than monarchs, aristocrats, and the very wealthy).

Before the marginalization of landscape gardening, a number of genres and forms of the art took shape, and Nietzsche shows a familiarity with many of the leading types in his books, notes, and letters. I will sketch a brief conspectus of some of the prominent European garden forms that Nietzsche knew directly, or from his readings, as in Burckhardt's *Cicerone*, a standard encyclopedic and critical guide to Italian art and architecture, which he frequently praised and recommended.[13] Given the focus of this study, it is worth noting that the garden has always had a political dimension. This is marked more strongly in the older forms, but is also implicit in newer ones. Let's then consider a few modes in which earth art discloses the sense or senses of the earth. All of these are highly but not exclusively visual constructions. There are many ways of describing these works, but I think it will be useful to adopt some of Foucault's archaeological analysis of the typical *epistemes* of the last six hundred years or so in articulating a few prominent sense-making genres. These gardens are enabled by what Foucault and Deleuze call diagrams, active arrangements of forces and materials; they are apparatuses for viewing and walking, involving all the senses.[14] In deploying these Foucauldian and Deleuzian concepts, I recall their own implicit and professed debts to Nietzsche in their formulations.[15]

The great Italian gardens of the Renaissance are governed by a diagram that proposes a correspondence, through analogy and similitude, of the microcosm and the macrocosm. The garden is not thought of as opening onto a free and unregulated nature but as exhibiting in miniature the structure and contents of a universal order. The prevailing ideals are plenitude and perfection, reinforced by astrological and mythological symbolism. It offers itself as a place of contemplation, its humanistic theorists drawing on the classical discourse of *otium*, arguing that such a place of retreat enables an intellectual intuition of order. Unlike the medieval cloister, however, the main direction of attention is horizontal rather than vertical, toward the plane of immanence, not the worship of a transcendent God. It stakes out a territory that offers an image or mirror of universal order.[16]

Very frequently, these gardens are not only theaters of elementals—water, atmosphere, earth (in the narrow sense), and even occasionally fire—a description plausibly applicable to almost all artistically ambitious gardens. They may also thematize this theatrical function. The Villa Lante garden (near Viterbo) was built in the later 1500s to accompany two grand houses. If most gardens are theaters of the elements, here water rules. Building upon the dramatic water designs of Persian gardens, transmitted to Europe through Moorish Spain, Villa Lante is a masterpiece of hydraulic engineering, which had recently become increasingly sophisticated. From the top of a hillside, water seems to spout naturally from the mouths of beasts or mythological figures rudely carved in stone. The waterspouts suggest a sense of primal origins. As water flows downhill, its architectural surround becomes increasingly polished and formed; the entire watercourse becomes an allegory of the progress of civilization in harnessing water for the ends of utility and beauty. Walking down the hill, following the water, you not only see it, but feel its cooling mist, hear its splashing and rippling, dipping your hands if you like. Finally, at the bottom, you contemplate a large, perfectly square pool adorned with stone figures and even stone boats, bordered by parterres of completely symmetrical clipped hedges. Water has become fully civilized. The meaning of the earth is its mastery by human civilization, which not only puts it to use but achieves a beautiful self-understanding of its vocation. The prince of the Church who commanded Villa Lante demonstrates both his own power and that of the family and institutions he represents.

Not far from Villa Lante is the garden of Bomarzo (known now as the Monster Park). Apparently constructed just a few years later, some think it was intended to upstage Villa Lante, in the constant cultural agon that accompanied the political and military conflicts of princes, cardinals, and city-states. Here the presiding element is earth. Bomarzo is studded with a number of chthonic beings, including serpent or dragon-like figures, symbolically tied to the earth and the underworld. Some have gigantic gaping maws that suggest entrances to the lower world. Death seems to have been very much on the mind of the landowner Francesco Orsini, a condottiere or mercenary soldier, who frequently faced and caused death. He began construction after his beloved wife died. We do not know whether he was expressing his grief, seeking his own avenue to Hades, or simply looking for a massive diversion. Chthonic forces assert themselves at Bomarzo, surging up out of the ground and opening up abysses. The site is now an amusement park with special appeal to children (of all ages), who savor the thrill of confronting their fantasy fears.

The classical French garden, for example in the work of André LeNotre at Versailles and Vaux-le-Vicomte, is not intended as a mirror of earthly nature; rather, these works are *Mirrors of Infinity*, as Allen Weiss calls this type in his eponymous book. It is a demonstration of sovereign power, of the ability to shape, structure, and master human and natural forces, clearly subordinating them to a central plan. Its territorialization is carried out by means of an emphatically rigorous and symmetrical design, so that the will of designer and sovereign is everywhere evident. The plan is geometrical; its articulation requires exquisite cartographical skills, and its execution draws on the developing science of engineering. There is a harmony of style in the French monarchy, the gardens of Versailles, and Cartesian philosophy. The underlying cosmology is heliocentric, replacing the earth as center with the sun, as the Sun King oversaw the construction of Versailles as a center from which to gaze into the infinite.

Oswald Spengler made a number of acute observations about the gardens of European modernity. Despite his hyperbole and questionable politics, Spengler should be read seriously, yet cautiously, for his accounts of comparative cultural spatiality. He emphasizes the modern garden's Faustian quest for unlimited views. A garden like Vaux-le-Vicomte, with its long linear borders, creates perspective lines that coincide only at infinity. Vision loses itself in the distances. For Spengler, this spatial directionality becomes simultaneously temporal, and is especially enhanced by autumnal accents of fading growth. In a grand and problematic stylistic parallelism, he extends the analysis beyond the garden:

> The late poetry of the withering garden avenues, the unending lines in the streets of a megalopolis, the ranks of pillars in a cathedral, the peak in a distant mountain-chain—all tell us that the depth-experience which constitutes our space-world for us is in the last analysis our inward certainty of a destiny, of a prescribed direction, of time, of the irrevocable.[17]

The English garden also participates in the temporalization of space, especially by its inclusion of ruins and in its fondness for reproductions of classical temples that establish a link to the past. As Spengler remarks, the ancients would never have imagined such an archaicizing interest. They knew their *Iliad* quite well, but could not have comprehended a quest to excavate the ruins of Troy.[18] The great garden at Stourhead is laid out so as to follow Virgil's account of Aeneas's travels, travails, and descent to the underworld, suggesting a complex political allegory. More generally, theorists of the picturesque like Uvedale Price prized garden elements that were weathered and showed the effects of time and age, from trees struck by lightning to peasant cottages in a visually pleasing state of mild decay.[19]

The informal garden aimed at a play of art and nature that could not be reduced to any simple pattern, instituting an ideal landscape that had been glimpsed before only in the paintings of Claude Lorrain, Nicolas Poussin, or Salvator Rosa. Its boundaries were disguised to suggest that it was a continuous part of the larger natural world. This was seen both by the English critics and their assiduous reader, Kant, as an art form of great significance. These gardens were the major form in which the art of the eighteenth century attempted to think of, and by means of, the earth.

The celebrated English gardens such as Stowe, Stourhead, and Alexander Pope's Twickenham, were constructed on philosophical lines, and their design involves a series of views and horizons that unfold in a tantalizing way so that the walker constantly meets new, unexpected views as she follows the garden path. These parks are typically dense with classical allusions that draw on the pastoral tradition of poetry and painting. The gardens included texts carved in stone as well as grottoes, follies, and, occasionally, "hermits" hired to establish the meditative tone of the site. John Dixon Hunt, one of the great garden historians, has perspicuously pointed out a significant change in the practice and aesthetics of garden design around the middle of the eighteenth century. The exemplary gardens of the century's first decades (Stowe, Stourhead) are symbolic and allegorical—they require a hermeneutics. Then philosophical (Lockean) empiricism replaced a culture of interpretation, and gardens were created for the taste of landowners who were not so firmly grounded in classical culture and who had less experience of the Grand Tour. At that point, interest shifted to more formal properties of the landscape. This was a shift in the meaning of the "picturesque." Originally the term was used to designate landscapes or other scenes that were worthy of history paintings—that is, paintings that revealed meaningful human actions. So the early eighteenth-century gardens told stories of British liberty, drawing on the humanist tradition. In just a few years, though, the "picturesque" tended to acquire its later meaning—Hunt calls it "vulgar"—in which it is the shape and disposition of the landscape that is crucial.[20] This is what British writers like William Gilpin and Uvedale Price meant by the picturesque. While they admired landscapes that were, they thought, like paintings (e.g., Claude), they dropped the interest in meaningful human action that characterized the earlier gardens and their aesthetic. It was apparently this later, "vulgar" sense of the picturesque that Kant absorbed, mostly through books and prints.

The English park or garden is the form taken by land art in the age of "man and his doubles." Here it is the later, "vulgar" picturesque that is most in keeping with the philosophical tendency of the time. It abandons pretensions to mirror cosmic order or to instantiate total sovereignty, just as the philosophy of finitude, whether in Kant, Marx, or the early Heidegger, gives up the aspirations of classical metaphysics to provide a rational, discursive account of being. From a cosmological perspective, it is Newtonian or Laplacean, accepting a decentered universe, privileging a multitude of apparently infinite horizonal perspectives. Yet the new form of the garden accepts the finitude of the viewer and stroller, who never achieves a single all- encompassing perspective.

The thinkers of the picturesque, like Gilpin, Uvedale Price, and Richard Payne Knight, theorize the effective diagram of the English garden as a structure of intricacy, complexity, and shifting perspectives at the same time that their contemporary Bentham was elaborating the diagram of the Panopticon as an instrument of total surveillance, including self-surveillence. Foucault calls the Panopticon "a machine for dissociating the see/being seen dyad."[21] If the Panopticon is the diagrammmatic form of the gaze at its extreme, the concretizing of Platonic vision and the inversion of the spectacle of the cave, with the spotlight now turned on the prisoners, the English garden, designed for those who regard themselves as very much at liberty, is the theater of the glance, the passing, perspectival, and partial look. In explaining the freedom of the imagination, Joseph Addison appeals to the experience of landscape and declares that "a spacious Horison is an Image of Liberty, where the Eye has Room to range abroad, to expatiate at large on the Immensity of its Views, and to lose itself amidst the Variety of Objects that offer themselves to its Observation. Such wide and undetermined Prospects are as pleasing to the Fancy, as the Speculations of Eternity or Infinitude are to the Understanding."[22]

So the landscape garden is a *machine for liberating the eye*, especially the eye of the landowner, whose view (and property) seem to be limited only by the horizon. To produce this effect, the English garden surrounds itself with a hidden frame, a ditch or earthwork known as a "ha-ha" (the term derived from military fortifications), invisible from within the garden, but which functions so as to exclude unwanted intruders, animal or human. This frame frames by concealing itself. It frames the territory by producing the illusion that there is no frame. Historians still argue over who first devised this crucial self-concealing contribution to artistic form. Horace Walpole saw William Kent as the genius who disclosed the amazing powers of the ha-ha in creating the classic landscape garden: he says that Kent "leaped the fence, and saw that all nature was a garden." The frame does its framing work by concealing itself. Jacques Derrida has explored this problematic status of the frame or *parergon* in Kant's third *Critique*. The English garden then exhibits an uncanniness that parallels the strange figure of "man," described by Foucault as an "enslaved sovereign, observed spectator." The frame is both internal and external to the park. It requires boundaries and limits and yet also must create the impression that it is continuous with the world. The English garden oscillates between a deterritorializing pictorial framing, as in Kant's optical reductionism or the image seen in the Claude glass, and an active reterritorialization that aspires to break down the boundaries between the world and the park. Later we will see that Nietzsche followed Burckhardt in voicing reservations over the sentimentality of the English garden.

Kant is generally recognized as marking a crucial stage in the formulation of modern aesthetics. Both landscape and garden play important roles in the argument of the *Critique of Judgment*, frequently invoked to illustrate the experience of "purposiveness without purpose," in which the faculties of imagination and understanding find themselves engaged in a lingering play. Flowers and gardens adorn the text of the third critique. Although the form was imitated in German lands, Kant may never have seen an "English garden." Nevertheless, he discusses this art form at several crucial points. Kant was aware of the work of many eighteenth-century British critics such as Burke and Addison, who did write about garden aesthetics; we don't know if he was acquainted with the writings of his exact contemporary William Gilpin (1724–1804), who contributed to displacing the word "picturesque" from its association with history paintings (and hermeneutic concerns) to designate a certain kind of landscape view, similar to the paintings of Claude Lorrain and others, in which human or historical meaning was of lesser interest.

In Kant's argument, we can see some intimations of the marginalization of garden and land art just a generation later. Landscape gardening is a form of visual art, and visual art is divided into arts of "sensible truth" or plastic art—sculpture and architecture—and "sensible illusion" or *painting*—including painting proper and landscape architecture.[23] We might wonder why landscape architecture is an art of illusion and not of truth, for the distinction seems to hinge on whether the products are offered to touch as well as to sight. But Kant says, "Landscape gardening consists in nothing but decorating the ground with the same diversity [of things] (grasses, flowers, shrubs, and trees, even bodies of water, hills and dales) with which nature exhibits it to our view, only arranged differently and commensurately with certain ideas."[24]

Notice that Kant makes no mention of allegorical or emblematic monuments, inscriptions, and the like, an indication of his having absorbed the "vulgar," posthermeneutic picturesque. More to the point, Kant's gardens and landscapes are beautiful only insofar as they remain objects of visual contemplation as given to the theoretical sense of sight (although aware of the charms of birdsong, Kant never acknowledges the possible combination of sight and hearing, the other theoretical sense, in the garden). Movement and desire have been drained from the garden experience, although they return in Nietzsche. Can't we—mustn't we—walk around and through a garden, paralleling our exploration of a building, taking its full three-dimensional ambience and atmosphere into account? The mobile glance of the stroller in the garden is different from the steady gaze directed at a picture on a wall. Doesn't a great park like Stowe invite at least as full a circumambulation and a sensitivity to mass as does Michelangelo's *David*? Kant's aesthetics is governed by a strong deterritorializing impulse, as he classifies an art that marks, shapes, and frames the earth as an art of illusion, in contradistinction to sculpture and architecture. At least three times in the section on the division of the arts, Kant assures us of painting's limitation to the eye. With no other art is he so insistent on its limits; note the apparently gratuitous "nothing but" in the preceding definition. Especially in the case of landscape gardening, Kant reiterates this restriction in a context suggesting that we might mistake the character of this art.

It was necessary for Kant to elide the history of the English garden in order to idealize it, and so reduce it to a purely optical form of painting. Consider now a more materialist, naturalist, or geophilosophical account of the form, one that sees it from the perspective of the *Menschen-Erde*. The genre was inspired in part by the paintings of Claude Lorrain and others, which wealthy landowners discovered on the Grand Tour. The pictures embody a pastoral ideal that was already an idealization of a landscape devastated by excessive grazing; too many herds eat up the low new growth of trees, and so, in a few generations, the forest disappears and we get shepherds in a landscape. This is an elementary form of *territorialization*, a largely unconscious form of environmental art. Paintings of the ideal landscape trade on the ability to transform this historical process into a timeless golden age, as in the paintings of Lorrain and Poussin that owe much to the humanized landscape of the Roman campagna. The paintings effectively *deterritorialize* the pastoral, making it available as an imaginary window on the collector's wall. Yet in a pragmatic spirit, the English travelers who brought these paintings back from their excursions treated them as sketches and designs for the transformation of actual swathes of land, moving the earth, planting and uprooting trees, diverting the course of streams, and building bridges, grottoes, and miniature classical temples. This was a *reterritorialization* of pastoral landscape in the rather different setting of the English countryside, and a constructive form of art, one that promiscuously lent itself to many deployments and variations.[25]

Nietzsche launches a withering critique of Kant's and Schopenhauer's aesthetics of disinterest in the *Genealogy*, tracing their divorce of desire and aesthetic experience to their commitments to the ascetic ideal. In concluding this polemic against the aesthetics of idealism, Nietzsche offers his own hypothesis about the roots of Schopenhauer's version and issues a promissory note concerning his own future aesthetics, one that unfortunately was never redeemed. The passage should be read in the light of Zarathustra's declaration that sensuality is "earth's garden-happiness." He puts forward the possibility that

> the peculiar sweetness and fullness characteristic of the aesthetic condition might have its origins precisely in the ingredient "sensuality" (just as that "idealism" characteristic of marriageable girls stems from the same source)—that sensuality is thus not suspended at the outset of the aesthetic condition, as Schopenhauer believed, but rather only transfigures itself and no longer enters consciousness as sexual stimulus. (I will return to this viewpoint at another time in connection with still more delicate problems of the thus far so untouched so unexplored *physiology of aesthetics*.) (*GM* III.8)

Nietzsche never fully developed this projected "physiology of aesthetics" (although perhaps a number of later aesthetic theories, associated, for example, with Freudian psychoanalysis or Deweyan pragmatism, can be seen as broadly consistent with his aspirations). Nietzsche's vivid accounts of various forms of garden-happiness should remind us that the project could be construed as not simply focusing on "internal" factors (from Burkean speculations about "fibers" to today's neuroaesthetics) but as taking its field of inquiry to involve humans situated in and responding to larger environments both natural and constructed. Aesthetics, as suggested by Sloterdijk and Gernot Böhme, would become atmospherics. From this perspective, the marginalization of landscape and garden in Hegel's aesthetics and its heirs can be understood as correlated both with idealism and the *Aufhebung* of geography by world-history. Gardens are major features of Nietzsche's earth but insignificant frames or interstices in Hegel's world.

Looking at Nietzsche's concern with gardens and landscapes from the perspective of the genealogy of aesthetics, we might note that he is one of several nineteenth-century thinkers who are eccentric in relation to the "mainstream" aesthetic tradition that had marginalized them. Here Nietzsche finds himself in the company of the untimely (and anti-Hegelian) Schopenhauer, who saw the will objectified in nature, and wrote of the inexhaustible fascination of fountains and falling water, whether constructed or natural.[26] The return of the repressed garden aesthetics is especially striking in Goethe's novel *Elective Affinities* (which Nietzsche knew and discussed with friends). The story revolves around a couple from the minor aristocracy, married relatively late in life, who find their marriage's chief mutual project in planning, laying out, and constructing a landscape garden in the English manner on their estate. Although typically (and rightly) described as one of the great novels of adultery, *Elective Affinities* interweaves this theme completely with the problematics of the garden as an aesthetic form. The tragedy of the initial couple is foreshadowed in their thoughtless treatment of the very grounds they are shaping and transforming, beginning with the wife's displacement of the established graveyard of the local church. They create a new lake from smaller ponds, but a dike collapses at its festive inauguration, leading to a near drowning. The tragic trajectory is confirmed when the couple's infant son (conceived while each fantasizes about another possible lover) actually drowns in the same lake. Throughout the novel, the reader sees that the builders disrespect their own natural and cultural surroundings. Rather than honoring Pope's "genius of the place," they willfully impose the pattern of the fashionable English garden on a German territory for which it is unsuited. The national aesthetic distinction is emphasized by their inattention to foundations, ground, and grounding, which are typically taken to be German concerns. Despite several warnings and hints from a few characters (e g., an architect and a widely traveled English connoisseur of landscape architecture), they persist in their fatal course. In *Elective Affinities*, Goethe provides an implicit critique and explanation of the couple's failure in their grasping at "earth's garden-happiness."

Garden aesthetics has a strong political dimension. This excursus into its history should help us to understand how Nietzsche could see the garden as an appropriate vehicle for conceiving of an earth in transformation, an earth seeking meaning and direction. Despite the more conventional temptation to see the garden as an escape from time, the European traditions of grand landscape architecture must be understood as projecting futures, not merely escaping from the present into a simulacrum of eternity. For Nietzsche, gardens are metaphors or symbols of a variety of activities, dispositions, and tastes. Yet more than metaphors, they are actually existing places, places that can promote the garden-happiness of the human-earth. The dominant themes are the shaping and tending of the natural, with a view to producing a rewarding result as well as the enjoyment of an earthly site.

ATMOSPHERICS AND PHENOMENOLOGY OF GARDENS

Gardens do not simply occupy space but are distinctive places (heterotopias, Foucault says) with specific atmospheres and invitations. Gardens, above all, have atmospheres, pervasive qualities that can nurture creativity. As Nietzsche writes in his *Unmodern Observation* on history: "Every living thing needs to be surrounded by an atmosphere, a mysterious circle of mist: if one robs it of this veil, if one condemns a religion, an art, a genius to orbit as a star without an atmosphere: then one should not wonder about its becoming withered, hard, and barren" (*UO* II.7). The "untimely," then, is not subject to the lockstep trajectory of modernity; the untimely is that which is situated in a place that provides a site for genuine "living." We might say that the untimely is the geographical and the geophilosophical. A garden, as Nietzsche describes it, is a heterotopia with "an atmosphere, a mysterious circle of mist." Let us assemble a few of the characteristics noted or implicit in our brief survey of the European garden tradition. Gardens are sites of becoming, responsive to variations in their surrounding context or environment. They are affectively significant and offer multisensory stimulation; they are atmospheric in a strong sense. Gardens maintain themselves on the threshold of nature and culture. They exhibit variations in form and genre comparable to those found in various arts and sciences. Gardens are experiments, *Versuche*, in living on the human-earth.

The concept of atmosphere can be helpful in clarifying the conjunction of site and futurity in Nietzsche's frequent invocation of the garden topos. Many phenomenologists make a sharp distinction between impersonal, measured, abstract *space* on the one hand, and *place* understood as a site with its own lived characters, qualities, paths, boundaries, borders, openings, and connections to other places. It should be clear that this kind of distinction resonates with Nietzsche's fear of the triumph of the last humans' totally measured world, with its standardized space and time, a world grown smaller and more uniform, no longer presenting a rich diversity of possibilities. Gardens have the potential for the cultivation of a variety of local forms of life and perhaps the cultivation of a richly productive future for the human-earth.

Building on the work of thinkers like Heidegger, Merleau-Ponty, Bachelard, and others, Edward Casey has explored the phenomenology of place in a spirit that complements Nietzsche's vision of the garden (or so I believe; Casey's references to Nietzsche are infrequent).[27] In Casey's diagnosis, place is highly specific and individual; it is richly complex and pregnant with possibilities of movement, experience, and action. Yet it has been overshadowed in the last few post-Cartesian centuries by abstract, universal space and time.[28] Especially interesting is Casey's emphasis on the mutual implication of bodies and sites in constructed places, including architecture and, importantly for our purposes, gardens, those products of landscape architecture: "In creating built places, we transform not only the local landscape but ourselves as subjects: body subjects become fabricating agents."[29] In his nuanced and highly articulated phenomenology of gardens, Casey identifies "three special lessons" that (I would argue) are implicit or explicit in Nietzsche's exploration of the topos. (1) Gardens "embody an unusually intimate connection between mood and built place." Where Casey speaks of mood, Nietzsche, the human barometer, might speak of "atmosphere," blending the meteorological or elemental sense of the term with the affective. (2) Gardens reveal "the expanded building potential of certain material elements," such as "ground, wood, water, and rocks." They are theaters of the elements in both the objective and subjective senses of the genitive. (3) Most importantly, Casey suggests, "gardens offer dwelling." They provide not only meaningful places to be but suggest further ways of acting in and on the earth. Like all dwelling, they involve both a certain stability and a ground of future movements, understanding, and transformation.[30]

Casey's work can be articulated with that of two recent theorists of atmosphere, Peter Sloterdijk and Gernot Böhme.[31] Sloterdijk's trilogy *Spheres* offers a rich analysis and typology of the structures—physical, cultural, fictional, and imaginary—in which humans both shape their dwelling on the earth and project further possibilities. For Sloterdijk, the crucial term to describe the emerging forms of human constructivist projects on the earth is "air conditioning." Over and above the apparatus or utility usually indicated by these words, he refers to the multifarious procedures that, whether deliberately or unconsciously, preserve and transform the territories or sites in which we live, work, play, with variously firm or porous borders, their texture and contexture indicated by his terms "bubbles," "globes," or "foam." Alteration may be destructive or constructive, ranging from Heidegger's fetishized peasant huts to the poison gas first released in World War I and subsequent acts of "atmoterrorism," including but not limited to aerial bombardment and drone warfare.

Nietzsche's sensitivity to atmosphere is also a question to philosophy, and not simply a consequence of his unique set of medical conditions. Why have philosophers had so little to say about atmosphere?[32] It is both ephemeral and inevitable, all encompassing but easily taken for granted. Philosophers, Nietzsche often reminds us, have difficulty thinking about what is right in front of their noses. Both Sloterdijk and Böhme stress that atmosphere cannot be reduced to a merely subjective, individual, idiosyncratic, pathological affect. They are attuned to a multiplicity of ways in which atmospheric affect can be artfully constructed. Böhme provocatively suggests that atmosphere can be taken as the key concept of a new aesthetic theory. He cites the practice of stage design, which has long been concerned with the production of atmosphere in the theater, a practice that has sometimes been the subject of explicit reflection by its practitioners, although almost completely ignored by philosophical aestheticians. To bring this observation into the current context, note that many of the great landscape architects (as we would now call these, garden designers) were either professional stage designers or learned from that art.[33] Now consider some of the ways in which Nietzsche articulates garden features, attending both to actually existing gardens and to certain extensions of the garden concept, either metaphorically or in his occasional speculations about large-scale transformations of the globe in order to promote "earth's garden-happiness."

SITES OF PLANTATION AND CULTIVATION

One obvious association of the garden is with marriage and sexuality. Speaking "On Child and Marriage." Zarathustra enjoins his listeners "Not only onward shall you propagate yourself, but upward! May the garden of marriage help you to do so!" (*Z* I.20; cf. *Z* III.12.24). Zarathustra speaks of the "trees" or children that he has planted and cultivated in his garden. "My children are still becoming green for me in their first spring, standing close together and shaken in common by the winds, the trees of my garden and my finest soil." He goes on to explain that as they mature, the "trees" will be transplanted, so that each can stand alone, weathering the trials of mountain and sea, becoming "gnarled and crooked and with pliable hardness" until, having passed such tests, it will one day become his companion. Here Zarathustra says that for now he must evade his happiness in order to continue this work of cultivation (*Z* III.3). Zarathustra's garden is a working garden.[34] Nietzsche sees the garden in these passages as a long-term experiment (some would say investment) in the future. Planting, and then transplanting to a more rigorous environment requires planning, time, patience, and sensitivity to the advantages and disadvantages of local circumstances.

PERSPECTIVAL POWER IN LANDSCAPE ARCHITECTURE

Nietzsche notes several times the desirability of fences and boundaries for one's garden. In some instances this has the function of protecting a space for contemplation and solitude, as when Zarathustra says, "I want to have fences around my thoughts and also even around my words, lest swine and swooners (*Schwärmer*) break into my gardens!—" (*Z* III.10.2). In this case, the garden is a garden of thoughts, which must be protected from naïve readings and appropriations (good luck with that, Fritz!). Yet Nietzsche makes the same point with respect to actual gardens. This indicates that he did some significant reading, thinking, and observation regarding various styles of landscape architecture; he was also capable of using the metaphor in the other direction, taking the style of actual gardens as indicative of possibilities of thought. In *Gay Science,* "Genoa" (291), Nietzsche draws such consequences from the earthly site where he began to compose *Zarathustra*. Here gardens are seen as landscape architecture, inseparable from architecture in a more limited sense; the architectural combination of house and garden involves a complex dynamic of the bounded and the boundless. Nietzsche begins by describing the human landscape: "I have been looking at this city for a long time, at its villas and pleasure-gardens and the wide circumference of its inhabited heights and slopes, and in the end I must say: I see *faces* that belong to past generations; this region is dotted with images of bold and autocratic human beings."[35] Here, then, is a region of the human-earth that has been inscribed with meaning. In reading the Genoa aphorism and related passages, it is helpful to visualize the actual topography Nietzsche describes. Genoa rises up steeply on several sides from its enormous bay; Nietzsche lived high on a hill from which he had magnificent panoramic views of much of the city. He must also have been at the shoreline and harbor from time to time, and so have looked up at the same slopes. The "faces" of the human landscape that Nietzsche sees are those of the builders—especially the city's maritime conquerors and explorers. Each of these "rests his gaze on everything built near and afar as well as on city, sea, and mountain contours . . . with his gaze he is perpetrating acts of violence and conquest: he wants to fit all this into *his* plan and finally make it his *possession* by incorporating it into his plan." The scene is one of vision and power. To see the gardens and houses is to imagine the views that each of them enjoys. What each builder wants is a comprehensive, commanding perspective; Nietzsche is no doubt thinking of the great palaces and gardens of the city, like that of Andrea Doria.[36] Moreover, we can say that these gardens *include* "city, sea, and mountain" insofar as they provide a frame for viewing them. They reach out to embrace their surroundings and thus create a distinctive atmosphere, the aesthetic equivalents of microclimates. The garden is a framing device for the intersection of the elements and the human.

These builders "heeded no boundaries in distant lands"; the attitude was translated into the urban setting as each, ignoring his neighbor, "flared up against the other and found a way to express his superiority and place his personal infinity between himself and his neighbor." Nietzsche explicitly contrasts this with the North's urban style that embodies lawfulness, obedience, and regularity. Yet "here you find upon turning every corner, a separate human being who knows the sea, adventure, and the Orient." Genoa's builders reach out to the farthest horizons; their houses and gardens aspire to unbounded, cosmic perspectives. Nietzsche's reasons for praising these gardens demonstrate some crucial aspects of his much-discussed perspectivism. Perspectives are not simply given by inheritance, local circumstances, or biology—although they may very well absorb or reshape any of these. Perspectives can be constructed, even though no one begins with a tabula rasa, just as they are by the many builders Nietzsche admires. Contrary to some hasty readings, Nietzsche's perspectivism does not imply that each perspective is monadically isolated from all others and can be maintained in total isolation from other perspectives. While it may be possible to double down on a single blinkered perspective, Nietzsche's perspectivism involves an awareness of competing alternatives. Rather, as with the multiple perspectives of Genoa's gardens and houses, each is conscious of a multiplicity of other perspectives and asserts itself within a perspectival *agon*. Nietzsche's perspectivism is an earthly perspectivism; it requires thinking with an *Erden-Kopf*.

In addition to the explicit contrast with the North, Nietzsche must be implicitly thinking of another contrast with the English natural or picturesque style of gardening. Nietzsche criticized the English taste in gardens as an exemplary *modern* taste, despite his love for the very painter, Claude Lorrain, whom the English took as one of their models. In notebooks of 1880–81, a period including stays in Genoa, Nietzsche makes several notes on the meaning of landscape and gardens. In the fall of 1880, he dips once more into Jacob Burckhardt's *Cicerone*, a companion to Italian art and architecture that accompanied him on his travels, leading to this notation:

> the taste of the *English* art of gardening—"to imitate free nature with its accidents" J.B.—is the entire modern taste. Such men wish to be poets: while it is another aim which those men have who "make the laws of art productive (*dienstbar*)." NB I must wean myself away from elegiac sentimentality for

nature. "The contrast of free nature, which shines into the Italian garden from outside" J.B. Fundamental condition of the impression. Such *men of style* work best within a *half wild* environment (*KSA* 9.255–56).

The passage in Burckhardt that Nietzsche cites and comments on celebrates the Italian garden as developed in the sixteenth and seventeenth centuries; he claims that with its great mastery of space and control of the intricacies of planting, avenues, line and perspective, fountains, and grottoes, it is impossible to imagine anything of this sort that would be more complete. He contrasts the weak, modern, English taste for "crooked paths, hermitages, Chinoiserie, straw huts, ruined castles, gothic chapels, and so on" with "the great, synoptic, symmetrical division of spaces with determinate character" of the Italian garden.[37] Burckhardt acknowledges, as reflected in Nietzsche's citation, that the Italian garden's effect is enhanced by the sight of free nature—mountains and sea coast, for example—beyond its bounds. The classic English picturesque garden of the eighteenth century, with many later imitators, eliminated visible architectural walls and fences, in order that the park might seem to blend seamlessly into the surrounding territory.

Nietzsche adopts Burckhardt's critique of the English garden as an egregious instance of modern sentimentality, understanding sentimentality as the determined project of blurring the boundaries of nature and art. While the garden is a complex assemblage of the natural and the cultural, it ought not to deliberately confuse the two in the English manner. Both the Italian and English styles rely on what Japanese gardeners call "borrowed scenery," but do so in quite distinct ways. The English try to produce the illusion of continuity and blending; the Italians forthrightly acknowledge their formative, constructive activity while reaching out to distant horizons. Nietzsche's garden aesthetic is consistent with his general philosophy of art, which stresses the power of framing, shaping, and making.

STYLES OF GARDENING AND OF SCIENCE

Nietzsche constructs an extended comparison between changing tastes in styles of gardening and landscape on the one hand, and approaches to science on the other (*D* 427). The aphorism shows that already in his "middle period" he was acquainted with the diversity and complexity of the European garden tradition, with its variations across time and national traditions. Rococo *Gartenkunst* arose from a distaste for nature seen as "ugly, savage, boring" and attempted to beautify it; just so, science aimed at its own beautification in order to be more entertaining, practicing a "deception of the eyes (with temples, distant prospects, grottos, mazes, waterfalls, to speak in metaphors), to present science in extract and with all kinds of strange and unexpected illuminations and to involve it in so much indefiniteness, irrationality and reverie that one can wander in it 'as in wild nature' and yet without effort or boredom . . . ." However, the beautifier of science should learn something from the change of taste with regard to the presentation of nature that dates from the time of Rousseau, when interest in the beauty of mountains and deserts (what Kant and others would call the sublime, although Nietzsche does not use the term here) replaced the prettified beauty of rococo taste. Nietzsche clearly understands that there is no normative style for the garden—or for science.[38] The corresponding shift in the scientific approach would find the highest intellectual beauty in the " 'wild, ugly' " aspects of its objects.

GARDENING AS A MODEL FOR THE CARE OF THE SELF

The garden is a model for the cultivation of the self. The aphorism "One thing is needful" is much commented on for identifying that one thing as "To give style to one's character—a great and rare art!" (*GS* 290). But what kind of art? This section immediately precedes the aphorism "Genoa" and draws on the notes Nietzsche made on Burckhardt and garden style; it describes two forms of success or failure in the project of self-styling; both are understood through the analogy of gardening or landscaping. I suggest that the question of what Nietzsche has in mind as an "art" by which style is produced is crucial for understanding this passage. Although the aphorism is much discussed, commentators are not always specific on this point. For example, in *Nietzsche: Life as Literature*, Alexander Nehamas takes Nietzsche to be at least implicitly invoking the model of literary art.[39] Moreover, Nehamas understands literary art as governed by the model of a totally unified "organic" work of art, in which every identifiable aspect of a work is an indispensable constituent of a meaningful whole. Whether Nietzsche actually endorsed such a strong normative aesthetic standard of organic unity in literature is open to question. More to the point is that literature is never mentioned in "One thing is needful," while architecture, landscape architecture, and interpretation of the natural environment (we could call it environmental aesthetics) are the only arts discussed in the aphorism.

A major difference between traditional literary art and the arts of building, gardening, and interpreting nature is that the successful literary work (as Nehamas assumes) attains a final completion and (organic) unity, while gardens are always in process, changing with seasons, weather, long-term climate change, local animal life, and countless other environmental factors. Nietzsche is obviously aware of this rather obvious dimension of gardens. Clearly, a garden will never be a completely stable and unified work. We can, however, distinguish types of gardens in terms of their most general stylistic features or lack of style (as Nietzsche does here and elsewhere). In "One thing is needful," Nietzsche first describes the person who "resists giving nature free rein," when "they have palaces to build and gardens to design." These artists of themselves take delight in dominating their materials and subjecting them to "a single taste" (here we should keep in mind what Nietzsche and Burckhardt say about the great Italian gardens). In contrast, "it is the weak characters with no power over themselves who *hate* the constraint of style . . . Such minds—and they may be of the first rank are always out to shape or interpret their environment [*Umgebungen*] as *free* nature—wild, arbitrary, disorderly, and surprising—and they are well advised to do so, because only thus do they please themselves!" This judgment echoes not only Burckhardt's remarks about the English garden but happens to agree (consciously or not) with Kant's aversion to some of the excesses he saw in the picturesque style.[40]

Nietzsche had elaborated similar themes in *Dawn*. In the aphorism "Gardener and garden," Nietzsche suggests, "Out of damp and gloomy days, out of loveless words directed at us, conclusions grow up in us like fungus: one morning they are there, we know not how, and they gaze upon us, morose and gray. Woe to the thinker who is not the gardener but only the soil of the plants that grow in him!" (*D* 382; cf. *D* 435). Similarly, "One can dispose of one's drives like a gardener, and, though few know it, cultivate the shoots of anger, pity, curiosity, vanity, as productively and profitably as a beautiful fruit tree on a trellis; one can do it with the good or bad taste of a gardener and, as it were, in the French or English or Dutch or Chinese fashion," or one can, with minor adjustments, "let the plants grow up and fight their fight out among themselves—indeed, one can take delight in such a wilderness" (*D* 560). In this aphorism and a few others, Nietzsche constructs corresponding analogies between styles of landscape painting and ways of structuring and representing aspects of one's own nature. The garden-like care of the self is not simply and narrowly selfish; it can be exemplary for others. We have a choice between suppressing the self in concern for others, which would lead to our becoming homogeneous grains of sand, or "creating something out of oneself that the other can behold with pleasure: a beautiful, restful, self-enclosed garden perhaps, with high walls against storms and the dust of the roadway but also a hospitable gate" (*D* 174). As in the "Genoa" aphorism of *Gay Science*, an individual's gardening—literal or metaphorical—is not a purely individualistic affair. It has strong effects on others, leading to a healthy agonistic culture. Conducted in the right spirit, gardening promotes a society of individuals with their own distinctive achievements, who can stimulate and welcome others.

The garden as constructive self-expression, cultivation for the future, perspective on the cosmos, and provocation for emulation can be paired and contrasted with the Epicurean garden as a site for care of the self and of temporary withdrawal, as required by circumstances. Nietzsche frequently invokes Epicurus's garden, where tradition informs us that he conducted his meditations and discussions with friends. The garden can be a place of refuge and relative solitude, providing congenial conditions for the philosopher's work, alone or with friends, and also a model for the care of the self. In *Beyond Good and Evil*, Nietzsche invokes the garden when warning the philosophers of the future against the dangers of martyrdom. They may find it necessary to flee or withdraw in order to accomplish their work. If so, they are admonished: "Don't forget the garden, the garden with golden trelliswork. And have people who are around you as a garden—or as music on the waters in the evening, when the day is turning into memories" (*BGE* 25). Stoicism and Epicureanism are contrasted as two forms of self-cultivation: the Stoic seeks out the most difficult terrain and conditions in order to harden himself, swallowing "stones and worms, glass shards and scorpions" in order to become insensitive to all external threats. This may be suitable for those who "live in violent times and depend on impulsive and changeable people" (Nietzsche might be thinking of the case of Seneca, who killed himself on the orders of his former pupil, Nero). But one who reasonably expects a relatively long life will do better by opting for the way of the Epicurean garden, providing a situation suitable for his "extremely sensitive intellectual constitution" (*GS* 306).

THE PROMISE OF HAPPINESS

When he relates his dream of weighing the world, Zarathustra transvalues the traditionally despised dispositions of sensuality, the lust to rule, and selfishness. There he redefines sensuality as, among other things, "the garden-happiness (*Garten-Glück*) of the earth, all futures' exuberance of thanks to the now" (*Z* III.10.2). It is this garden-happiness that Zarathustra feels free to enjoy, as he convalesces, upon his animals' invitation. When Nietzsche expresses admiration for Epicurus and his "garden philosophy," he emphasizes the importance of the garden as a protected refuge (in addition to active care of the self).[41] So far as the garden is a work of art, it exemplifies Stendhal's saying that art is a "*promesse de bonheur*," a position that Nietzsche opposes to the Kantian-Schopenhauerian theory that art and the aesthetic are forms of disinterested selfless contemplation (*GM* III.6). Readers will recall that Nietzsche's published texts, his letters, and his notes are strewn with descriptions of his heightened, enticing experiences of landscape and scenery. Especially suggestive is "*Et in Arcadia ego*" (*WS* 295), which portrays an Alpine landscape as mediated by the paintings of "Poussin and his pupil" (probably Claude Lorrain). The title does double duty: it is both that of two of Poussin's celebrated paintings and the epigraph of Goethe's *Italian Journey*. Nietzsche's description of the scene is painterly, emphasizing its composition. This makes it garden-like in several ways: as an agricultural landscape, it has already been humanly organized, and the painters Nietzsche cites are those who were notably influential in stimulating later garden design.[42]

THE GREAT TREE OF HUMANITY

An intriguing but necessarily sketchy outline of what the earth might be like as a garden appears in a sequence of two sections of *The Wanderer and His Shadow* (188–89). Here Nietzsche, however schematically, approaches the idea of an environmental politics for a garden earth. In the first, "Spiritual and physical transplantation as remedy," Nietzsche outlines a "medicinal geography" or "pharmacology," which would "send each person to the climate favorable to him—for a period of time or forever." Each human culture can be considered as a "spiritual climate" that provides what may be nurturing or harmful conditions for specific "organisms," obviously including humans. Nietzsche sometimes described his task as that of a cultural physician; here such a doctor would prescribe relocation to individuals who are likely to flourish best in an alternative physical and spiritual climate. This is only the outline of a new discipline, for much still has to be learned about the advantages and disadvantages of the various sections of the earth for people of different constitutions. As such knowledge accumulates, "nations, families and individuals must be gradually transplanted for as long and continuously as is needed for our inherited physical infirmities to be conquered. In the end, the whole earth will be a collection of health resorts" (*WS* 188). In other words, the cultural physician or pharmacologist of the future will see the earth as a resource for promoting the "garden-happiness" of groups and individuals.

In the following section, "Reason and the Tree of Humanity," Nietzsche reveals that (like virtually all his contemporaries) he has no worries about exhausting the finite resources of the earth.[43] While some, in "senile shortsightedness," fear that the earth is becoming overpopulated, he says, it is this profusion that "proffers the more hopeful their greatest task," namely, to reshape the earth as the ground of a great tree of humanity. Following Nietzsche's image of a tree that "overshadows the whole earth," and perhaps literalizing it a bit, the suggestion seems to be that future humanity, in its many billions, can grow upward and outward, relying on the nurturing ground and resources of the earth (here understood specifically as the globe). He realizes that the task he imagines is "unspeakably great and bold." In a rather different tone than the preceding section, which seems to promise health resorts for all, Nietzsche here envisions the cultivation of the "great tree" as an experimental project destined to last centuries and involving the suffering of "individuals, nations and ages." However, the goal is "the great collective fruit tree of humanity." This picture begins to sound like an exaggerated version of Sloterdijk's vision of an air-conditioned world of foamy and porous cells. Certainly there is much to question in this call to centuries of widespread sacrifice in order to reach a distant, relatively general goal through admittedly risky and as yet unformulated means. At least it can be said that Nietzsche does not minimize the risks. Yet the extravagant image of the great tree clarifies several aspects of "garden-happiness." The great garden, perhaps better seen as a multiplicity of gardens, must be experimentally constructed. Such gardens will be differentiated according to varying needs of individuals and groups, making maximal use of earth's resources for the sake of human flourishing. A related image of upward growth in *Beyond Good and Evil* is explicitly aristocratic. There Nietzsche compares the situation of a "good and healthy aristocracy" to that of "the sun-seeking vines of Java—they are called *Sipo Matador*—that so long and so often enclasp an oak tree with their tendrils until eventually, high above it but supported by it, they can unfold their crowns in the open light and display their happiness" (*BGE* 258). The earlier "great tree" image not only places more emphasis on the collective good of the many billions who will flourish on it; it also differs from this analogy of the vines insofar as it involves cultivation and planned experiment rather than a wild or merely natural growth. While the analogy of the Javanese vines is *only* an analogy between vegetative growth and the emergence of a happy aristocracy, the discussion of the great tree explicitly speaks of transforming the earth.

Nietzsche's implicit assumption that humans will not destroy or exhaust the earth's resources is expressed poetically in the *Zarathustra* chapter "On Great Events." There Zarathustra declares that despite humans constituting a disease on the "skin" (surface) of the earth, "the heart of the earth is of gold" (*Z* II.19). This suggests the need for transformation and healing of humanity as a growth on the earth—in rather sharper terms than *WS* 188–89—while also offering an unsupported faith in earth's potential for self-renewal. Today Zarathustra's account of earth's skin, its diseases, and its heart could be read as consistent with the "Gaia hypothesis," which sees the earth as a single living organism. At the end of the "great tree" passage, Nietzsche describes the enormous project of preparing the earth as "a task for reason on behalf of reason!" It is a task for reason, we might say, to cultivate the earth with a view to human flourishing; that is, (experimental) reason must be employed in the transformation from haphazard growth to giant garden. It is also a task on behalf of reason, especially when reason is taken in the larger sense that Nietzsche frequently gives it, of the "greater reason" of the (spiritualized) body, the body that no longer "buries its head in the sand of heavenly things, but carries it freely, an earthen head (*Erden-Kopf*) that creates a sense (*Sinn*) for the earth!" (*Z* I.3). It is especially important to emphasize that reason is understood here as an experimental enterprise, rather than in terms of unchangeable formulas; Nietzsche envisions that the good gardeners of humanity are capable of learning from their experience. Recall here his frequent invocation of Epicurus's philosophical garden. The garden is not only a place of refuge but an incentive and possible model for thinking of the human-earth as a site of garden-happiness.

Especially in *Dawn* and *Gay Science* it becomes clear that Nietzsche's talk of gardens, urban design, and landscaping is more than metaphorical. This is especially the case in Book Four of *Gay Science*, which concludes with the first explicit discussion of eternal recurrence and the introduction of the figure of Zarathustra. We might say that the teaching and the teacher are already set in landscapes that anticipate the teacher's earthly itinerary and the garden that discloses itself through his teaching. Nietzsche muses on the design of the urban environment when he writes of an "*Architecture for those who wish to pursue knowledge*":

> One day, and probably soon, we will need some recognition of what is missing primarily in our big cities: quiet, and wide, expansive places for reflection—places with long, high-ceilinged arcades for bad or all-too-sunny weather, where no shadows or noise from carriages can penetrate and where refined manners would prohibit even priests from praying aloud: a whole complex of buildings and sites that would give expression to the sublimity of contemplation and stepping aside.

After explaining why church spaces cannot allow "us godless ones to be able to think our thoughts here," he issues this design ordinance: "We want to have *us* translated into stone and plants; we want to take walks *in us* when we stroll through these hallways and gardens" (GS 280).

GARDENS AND VOLCANOES

There is an obvious tension between these two earthly sites: the volcano where Zarathustra discourses on great events, affirming earth's power of self-renewal, and the garden that Nietzsche projects as the form of earth's future. The hazard of destruction and death lurking in earth's molten core (its "heart of gold") seem to threaten the "garden-happiness" or the "great tree of humanity" that Nietzsche imagines as earth's direction. Nevertheless, he sees this as a productive tension rather than an impasse. As Nietzsche famously says, "Believe me—the secret for harvesting from existence the greatest fruitfulness and the greatest enjoyment is—*to live dangerously!* Build your cities on the slopes of Vesuvius!" (*GS* 283). Earthquakes and volcanoes are fairly constant elements of Nietzschean geography. While others typically emphasize their terrors, he tends to think of them as stimulating and invigorating. Never far from Nietzsche's thought is the iconic figure of Empedocles, whose legendary descent into Aetna provoked him (following Hölderlin's example) to attempt several dramatic versions of the tragic philosopher and his fate.[44] An aphorism in *Human, All-Too-Human* suggests that volcanoes are not only threats but sources of nourishment for gardens. The richness of volcanic soil makes for luxuriant growth, as crises on the human-earth promote new forms of happiness. Entitled "Vegetation of Happiness," the section articulates the relation between convulsive upheaval and the production of grounds for happiness on both literal and metaphorical levels:

> Right beside the sorrow of the world and often upon its volcanic ground, human beings have laid out their little gardens of happiness; whether they observe life with the eye of someone who wants knowledge alone from existence, or of someone who submits and resigns himself, or of someone who rejoices in difficulties that have been overcome—everywhere they will find some happiness sprouting beside the misfortune—and indeed, the more happiness, the more volcanic the ground was—only it would be ridiculous to say that the suffering itself could be justified by this happiness (*HAH* 591).

As the last sentence suggests, Nietzsche is aware here of the dangers of a reductive naturalism that legitimates destruction by its fortunate sequel, even if his later, more explicitly tragic, philosophy verges on just such a move.

Nietzsche's idea of a complex conjunction of volcano and garden suggests that the ravages of nature are a necessary price to pay for the glories and abundance we can cultivate in their aftermath. For most of human history, this is at least an initially plausible thesis. It can be illustrated by some of those cities and gardens built on the slopes of Vesuvius, or by a visit to the tropical gardens and rainforest parks on Hawaii's volcanic islands. Yet now the price has risen exponentially, with the development of technology on a scale unknown in the 1880s. Industrial accidents and environmental degradation, triggered either by natural events like earthquakes and tsunamis, simple human error, or corporate and governmental negligence have raised the stakes enormously. Nuclear disasters, releases of massive amounts of toxic chemicals, the ruin of the oceans, climate change—these and related dangers cast doubt on the relative tolerance that Nietzsche suggested we should extend to earth's convulsions.

Not surprisingly, a number of post-Nietzschean philosophers have wrestled with the problems of hypertechnology. Peter Sloterdijk analyzes the twentieth century and beyond as the time of *Terror from the Air.* Beginning with the first military uses of noxious gasses in colonial wars, through aerial bombing (already practiced before 1914), to Hiroshima and Nagasaki, to the constant threat of nuclear, chemical, and biological war, and finally the acceleration of climate change—all these have transformed the image of the protective heavens (such as Zarathustra praises it in "Before Sunrise") into an unpredictable site of attack or of "airborne toxic events" (Don DeLillo's phrase). In *After Fukushima*, Jean-Luc Nancy meditates on this change in our relation to the earth, a relation in which a tsunami can now cause waves of devastation that were impossible earlier, and that now seem at least periodically inevitable.[45] Nancy could be read as agreeing with Nietzsche's analysis of what the last humans are doing to the earth. The need for ever-greater security, control of risks, and providing assured means of sustenance and comfort require constantly increased metameasures and new procedures of regularization, all of which push the project of industrialized humanity and capitalism into new reaches of risk.

EARTH'S GARDEN-HAPPINESS

Nietzsche's vision of a garden earth is a radical alternative to both the petty politics of the nineteenth-century nation-state and to the loathsome prospect of a leveling down to the last humans' self-satisfied mediocrity. As opposed to nationalism, with its wars, superficial enthusiasms of the multitude, fetishism of soil and ethnicity, and emerging tyrants, the garden and the great tree of humanity promise to free us from the "shadow of God" embodied by the state.

They are not only sites of rest and convalescence, but projects requiring cultivation and design, opening to a future on the earth rather than accepting the dead ends of world-history. Yet however salutary, invigorating, and promising the vision, it seems to remain only a vision, a fantasy conjured by Nietzsche, perhaps, in his constant quest for ideal climates and surroundings, which led him to seek out choice sites in the Swiss Engadin, the French Riviera, and a series of Italian cities and resorts. Nietzsche's projection of a possible garden-happiness of the earth should be seen as both an aestheticized politics and a politically inflected aesthetic. But for the "world" to truly move toward becoming a garden earth or great tree, it must cast off the political theology that has empowered the West and its very conception of the world.

CHAPTER 6

Earth, World, Antichrist: Nietzsche after Political Theology

If there is one point where one realizes that Christianity has twisted humanity, it is to be found in all the connections that Christianity has entered into with politics, just as I do not doubt at all that here is the point at which it will one day succumb to a general contempt.

FRANZ OVERBECK, letter to Heinrich Treitschke

Would you like a new name for me? The church language [*Kirchensprache*] has one—the *Antichrist*. Let's not forget to laugh!

NIETZSCHE, letter to Malwida von Meysenbug, April 4, 1883

It is certain that the political philosophy of modernity will not be able to free itself of its contradictions if it does not become aware of its theological roots.

GIORGIO AGAMBEN[1]

KINGDOMS OF HEAVEN AND EARTH

Perhaps our future involves a network of glorious gardens or a great green rhizomatic outgrowth of the earth, rather than the shrunken, measured world of the last humans. The human, that Zarathustra once describes as earth's skin disease or autoimmune infection (*Z* II.18), may yet experience an *Umwertung aller Werte* that will enable such transformations. Yet as we have just seen, the garden earth remains a hope, possibility, and promise in Nietzsche's thought. The "perhaps," as Derrida reminds us, is an insistent dimension of this thinking.[2] Still, the invocation of the figure, metaphor, or phantasm of the garden forms, we might say, the atmosphere within which this most atmospheric of thinkers engages with his time and with eternity.

Today it is difficult but not impossible to sustain Nietzsche's promise that some future human will restore "to earth its purpose and to humans their hope"

(*GM* II.24). The technology, observation, and industry that reveal the earth as a single biosphere are deeply complicit in the worst eruptions of the anthropocene skin disease. The culture shows an insatiable desire for apocalyptic films, fictions, and prophecies of the world's end or earth's ultimate catastrophe. In addition to imagined cosmic dangers (sun eruptions, comets striking the planet, predatory or infectious extraterrestrials), transformations triggered by human industry or conspiracy express current fears (nuclear war, disastrous global climate change, unprecedented epidemics, biological engineering gone wrong, or our own machines waging war against us). A few movies or novels imagine a possible utopian solution or at least the survival of a remnant with the opportunity for a fresh start (the types can be combined). The first type yields extremes like Lars von Trier's *Melancholia* (total destruction of earth, and the world with it) or *The Road* (Cormac McCarthy's novel, John Hillcoat's film of abysmally grim postapocalyptic life and death). Christopher Nolan's film *Interstellar* and Kim Stanley Robinson's science fiction novel *2312* present another option in which technology is reborn as redeemer. Here space travel has allowed the creation of new extraterrestrial enclosed, self-sustaining green environments, complete with adjustable gravity, and medically enabled life extension. While these apocalyptic narratives often draw on themes analogous to those in Revelation, the ultimate Biblical story of the last days, they typically do not thematize the theological parallels. In *The Day After Tomorrow*, the Gutenberg Bible is saved because it is a testimony to human inventiveness and the age of mechanical reproduction, not for its religious value.[3] There are also, of course, a great number of stories and prophecies throughout the media with much more explicitly religious themes, often featuring a countdown to Armageddon or the rapture, as well as identifications of contemporary figures with the Beast, the Antichrist, and the Whore of Babylon.

We could read all these apocalyptic and postapocalyptic visions as commentaries on both Zarathustra's scorn of the earthly human "skin disease" and on his maxim that "the heart of the earth is of gold." We might expect that Nietzsche's earth of the future, still and always to be brought into being, would be free of the otherworldliness that has squandered earth's potential. As Zarathustra says, summarizing his political program to the Higher Humans, including a pair of kings, the last pope, and God's murderer, " 'we do not want the Kingdom of Heaven at all: we have become men—*and so we want the Kingdom of Earth (Erdenreich)*' " (*Z* IV.18).[4] Yet Nietzsche rejects secularization as an illusion, borrows or adapts Christian symbolism in projecting "a philosophy of the Antichrist," and finally identifies the entire project of transvaluation with his ominously titled book *The Antichrist*. It is time now to interrogate this continuing engagement with Christianity in terms of his rethinking of earth, its futurity, and its multiple times and temporalities. Nietzsche's friend Franz Overbeck (whose work, I argue, contributed heavily to *The Antichrist*'s polemic) had grave reservations about the text, which he nevertheless carefully transcribed for himself. In the circumstances (having retrieved Nietzsche from Turin after his breakdown), we can understand Overbeck focusing on the book's seeming madness. Today it is time to read Nietzsche's summoning up the paradigmatic Western story of how the world comes to an end, who shall rule the earth, and the accompanying disruptions of times and timing. This reading should prescind, if possible, from the biographical obsessions that usually dominate responses to Nietzsche's late writings.

If we read Nietzsche as rethinking and naturalizing the Biblical thought of paradise, it may not be so surprising that he became increasingly concerned to confront the end of the religious story that begins in the garden. Does the recovery or reconstitution of the heterotopic garden as a site of unmortgaged becoming (*Unschuld des Werdens*) require a final settling of accounts with the Christian metanarrative of the world's apocalyptic end? To lay the groundwork for restoring the earth must we rewrite the traditional story of the end-times and their fearsome symbol, the Antichrist?

Zarathustra's apparently simple dichotomy of the kingdom of heaven and the kingdom of earth in his address to the Higher Humans might appear to be a candidate for the deconstruction indicated in Agamben's observation, concluding his essay on Hobbes, that the political philosophy of modernity must come to terms with its theological roots. Is the earthly kingdom just a secularized version of the heavenly one? Agamben is impressed by Schmitt's thesis that all modern political concepts are secularized versions of theological ones. In fact—although Agamben does not dwell on this—Hobbes himself was able to think of the "Kingdom of God" as an earthly kingdom, a civil kingdom on the earth, that (as he reconstructs Biblical narrative) lasted from earth's creation to the kingship of Saul.[5] It has been interrupted ever since, given over to other forms of rule—ideally the absolute state or Leviathan—and will be restored on earth when Jesus returns to reestablish it in perpetuity. (Agamben notes that despite *Leviathan* focusing explicitly in its second half on the theological and the hermeneutic, Hobbes evades a number of difficulties by not mentioning the medieval identification of his titular beast and Antichrist.)

Nietzsche, like Agamben, is skeptical about whether so-called secularization is a truly transformative process. We should not assume that Zarathustra's earthly kingdom is a naïve reversal of the heavenly one. The concept "kingdom of God" and all it supposes or implies must be carefully analyzed in order to avoid imposing its character on any new conception of human arrangements. We hear a similar caution and warning from the madman, who says that those who simply scoff at God's existence have no sense of the consequences, magnitude, and agency of his death-event (or our complicity in his murder). For example, as Nietzsche's text suggests, they may (like Schmitt) endow the state or (like globalizing neoliberalism) the market with the absolute power previously reserved for God (*GS* 125). Similarly, we could say that Hobbes works the analogy in a reverse direction, arguing from the rationality of absolute sovereignty in our fallen era to the structure of the coming civil kingdom of Christ. Perhaps we are still waiting for the full news of God's death to reach us, a message that arrives slowly, like the light from distant stars. When this cosmic code is finally decoded on our planet of clever animals, we will cleanse our caves further of the lingering shadows of God (like state and world market).

Until this great event, Nietzsche offers us a twofold approach. On the one hand, he counters religion's idealistic self-understanding as emanating from a world-beyond (*Hinterwelt*) with a natural history, showing that all religion is bound to the earth, even when cursing it or denying its reality. On the other, he challenges the specific theo-logic and sacred narrative of Christianity by which it rationalizes its contempt for the earth. The first strain in Nietzsche's thought is important, but since it has received a great deal of attention by many scholars over the years, I will focus on the second, which Nietzsche sometimes called "a philosophy of the Antichrist."

This is a project that Nietzsche never finished, although his last substantive book is titled *The Antichrist*. In various notes and letters of his last months, that book was at first said to be the first of four in the planned *Hauptwerk*, *The Transvaluation of Values*, then said to be the whole of it. Given his rapidly changing program during a chaotic time, we are entitled to doubt whether the *Transvaluation* was completed; it is even more certain that he never wrote a book with the title *Will to Power*, another project title that has led to myriad misunderstandings because of editors' mischief. Yet we can ask what "a philosophy of the Antichrist" might look like. What is this philosophy that speaks with a name, that names itself? I propose to explore that question in several stages. First, I want to clarify Nietzsche's reasons for rejecting the idea of secularization as a political illusion; that rejection sets the stage for considering how and why he continued to engage—antagonistically, to be sure—with the Christian tradition. Second, I give an account of Nietzsche's familiarity with and understanding of Christian political theology, specifically including the figure of Antichrist. This can be considerably clarified by drawing on the research of his friend and colleague Franz Overbeck, the Basel theology professor and "comrade in arms," with whom Nietzsche maintained a regular correspondence after leaving the university and house where they both had apartments. Finally, I make some suggestions about how Nietzsche's admittedly fragmentary and sketchy "philosophy of the Antichrist" might contribute to the thematic complex of earth, great events, and great politics, or as he says in the *Genealogy*, how the coming Antichrist could restore a future to the earth and hope to humanity.

BEYOND SECULARIZATION

The question inevitably arises why Nietzsche, the advocate of radical immanence, would inscribe himself into Christian tradition—even in the form of outrageous polemic—by evoking its themes in "a philosophy of the Antichrist." Why would a text with the name of this apocalyptic figure, the focus of religious paranoia since the early Church Fathers, be designated as either the first volume or the whole of the projected *Transvaluation of All Values*? Why honor the tradition by taking it seriously? Why not stay within a strictly secular orientation? How does this sophomoric diatribe contribute to Nietzsche's project of "giving back its goal to earth and hope to humans" (*GM* II.24)?

Perhaps Nietzsche realizes that simply attempting to deny or evade the theological tradition risks unintentionally reinstating it in other terms? Here we need to see that Nietzsche is no friend of secularization, and regards it as a political delusion. What did he mean by secularization? The German term here is *Verweltlichung*, and it has a history that needs to be unearthed. First, this term could be translated more literally, if awkwardly, as "worldification," a process of making all life worldly. Yet as we have already seen, Nietzsche regards "world" as a theological concept, whether taken (as in earliest Christianity) as a realm of sin to be escaped or, in modern thought (e.g., Hegel) as the correlative of that shadow of God, the state.

Carl Schmitt offers a useful provocation here. His *Political Theology* claims that secularization is parasitic upon theology:

> All significant concepts of modern political thought are secularizations of theological concepts not only because of their historical development—in which they were transferred from theology to the theory of the state, whereby for example the omnipotent God became the omnipotent lawgiver—but also because of their systematic structure, the recognition of which is necessary for a sociological consideration of these concepts. The exception in jurisprudence is analogous to the miracle in theology.[6]

It is this "secret signature" of secularization, as Agamben says, that can be deciphered to show theology manifesting itself under a different name. Although he seems to side with Schmitt against Max Weber, who defended a more "innocent" thesis concerning modern political secularization, Agamben wants to recall not only the theological genealogy of sovereign authority (as Schmitt does) but to make explicit the economy (*oikonomia*), related to Foucault's concept of governmentality, that was elaborated by the early Church Fathers.[7]

Hegel seems to confirm the general lines of Schmitt's and Agamben's understanding of political theory when, in his *Philosophy of World History*, he robustly defends modern secularization and at the same time says that we are living in the last days, in the true Christian understanding of that idea. In introducing his analysis of the Germanic world, Hegel announces that "the Christian world is the world of completion; the grand principle of being is realized, consequently the end of days is fully come."[8] Now, in the end of days, it turns out, "secular life (*Weltlichkeit*) is the positive and definite embodiment of the spiritual kingdom—the kingdom of the will manifesting itself in outward existence."[9] In this absolute Protestantism, with its construction of the nation-state and its complete *Aufhebung* of the earth (or human geography) in world-history, Hegel achieves a politico-theological synthesis whose only glimpses of futurity are incidental and tentative speculations about which states will rise and fall.

We can read Nietzsche's ironic distancing of himself from the "political delusion" of secularization in a note from 1881, written just after his first notes on eternal recurrence. We should recall that these notes include many suggestions concerning the political relevance of the new teaching.

> The political delusion at which I smile in just the same way my contemporaries smile at the religious delusion of earlier ages, is primarily secularization (*Verweltlichung*), belief in the world (*Welt*) and a deliberate ignoring of the "beyond" and the "afterworld." Its goal is the well-being of the ***fleeting*** individual: which is why its fruit is socialism, i.e., fleeting ***individuals*** want to conquer their happiness through socialization—they have no reason to wait, as do human beings with eternal souls and eternal becoming and future improvement (*KSA* 9.504–05).

Nietzsche construes secularization in two political registers: as a *concern* for all "fleeting individuals," which necessarily reduces them to their lowest common denominator, and a *legitimation* of a strong central authority to guarantee standards of welfare. Secularization, at least in its late nineteenth-century form, is a theological delusion, insofar as it reduces all humans to the same level (comparable to Christianity's teaching that we are all equal children of God) and that a single central power must be instituted to provide for them (as the Church claimed to oversee their possibility of salvation).[10] In other words, the "illusion" of secularization in the nineteenth century involves both sovereignty and economy. In *BGE*'s "Peoples and Fatherlands," as we have seen, Nietzsche suggests that a certain level of homogenization and mobility, that is, a situation in which individuals see themselves as "fleeting individuals," is fertile ground for the production of "tyrants of all sorts" (*BGE* 242).

The tendency of Nietzsche's analysis of secularization, then, is deconstructive. He demonstrates that the hasty attempt to negate a metaphysical position typically results in simply reproducing its outlines in altered form, as if a photograph were replaced by its negative. It is, of course, these "secret signatures" or "shadows of God" that must be exposed and undermined from within through a thought that dares to speak of "a philosophy of the Antichrist."

An exemplary form of such deconstructive analysis appears in the much-discussed parable "How the True World Became a Fable," whose final thesis is that, contrary to the assumptions of the tradition, the critique and implosion of the concept of the true world does not leave us with an apparent world: that apparent world is now revealed to have been simply the foil to the true world. The very category of "world" has disappeared. This brilliantly condensed deconstructive history of philosophy since Plato focuses on *Welt* or world. The later senses of this concept are absent in Plato, who knows of the *kosmos* and *gaia*, but not of the world in the sense that *oikoumene* begins to acquire with the Stoics, as the entire realm of (known) human habitation, or in the New Testament's suspicion of the wisdom and customs of the *kosmos*.[11]

IT'S LATER THAN YOU THINK

In his last days before the breakdown, Nietzsche imagined that *The Antichrist* would be published simultaneously in seven languages, in editions of at least one million each. He expected to make an impression. Was it to herald a great event or to be that great event itself? One thing we can say with confidence is that Nietzsche is declaring that things are rapidly coming to a head, a threatening, game-changing head, much more quickly than anyone suspected. Great events are not predictable and arrive unexpectedly, sometimes "on dove's feet." So far, Nietzsche's literary tactic coincides with the formal core of the apocalyptic film in which (as in *The Day After Tomorrow*) the earth is cracking, the continents moving, the new ice age arriving in suddenly cascading sequences. Even as new computer modelings are produced, the events overtake them. Those who prognosticate from Revelation based on the signs of the times breathlessly report that we are already very far along as everything rushes to a conclusion. In 1888 the Antichrist may have seemed like a rather quaint figure to much of Nietzsche's potential audience, but one thing the legend stood for was a sudden, massive shakeup of everything. Before unearthing other aspects of the Antichrist figure, let us note that Nietzsche's text *The Antichrist* (and the "philosophy of the Antichrist") clearly signals its concern with questions of time and temporality from beginning to end. This text not only observes and analyzes a plurality of temporal modes, but aims at actively intervening in its readers' sense of time. It demolishes core Western narratives by portraying an atemporal Jesus, explaining Christianity as Paul's political invention, and offering heterodox stories of Christianity's relation to Rome, Islam, the Crusades, and Germany. From outside, from a distance suggesting a parallel universe with its own time, "we Hyperboreans" (we've somehow been incorporated in the "very few") become temporal guerillas, disrupting and interrupting world-history, splitting it in two. Here we might further gloss Agamben's remark that "it is certain that the political philosophy of modernity will not be able to free itself of its contradictions if it does not become aware of its theological roots." It is the political philosophy *of modernity* that's at stake. Modernity is a self-named, self-described time. To be modern is to be able to place classical antiquity, the Middle Ages, Renaissance. To be modern is to be up-to- date, and Nietzsche wants to disrupt all dates, emblematically by starting a new calendar with *The Antichrist*. Nietzsche was squaring off against such modern thinkers as Hegel, Comte, Spencer, and Darwin. What makes them modern is their fundamental commitment to the idea of "progress," the notion that history, society, or biological life are developing, evolving, or unfolding in a movement toward greater complexity. These modern thinkers tell stories of fulfillment and realization that owe much to Christian metanarrative, despite their self-described "secularism."

If "we Hyperboreans" are not modern, what is our time? When are we? *The Antichrist* suggests a dizzying variety of answers. In other words, it situates itself in a plurality of times. The Foreword repeatedly distances itself from the readers and culture of "today" and from today's "ephemeral chatter (*Zeitgschwätz*)." "Only the day after tomorrow belongs to me," Nietzsche intones. Tomorrow is on the horizon, we begin to see our way there, but the day after tomorrow is further away; it is less known and predictable. Time may fall out of joint. The "modern human," we then hear, is unhappy and confused, sighing, "I know not which way to turn." This modern human may be bewildered by the many different versions of the cult of progress propagated by modern thinkers from Hegel to Darwin. These thinkers of progress tell us that humankind is improving, but "mankind does not represent a development of the better or the stronger or the higher in the way that is believed today. 'Progress [*Fortschritt*]' is merely a modern idea, that is to say a false idea." Higher types of humans are possible, but not through modern progress. Rather, such "happy chances [*Glücksfälle*]" have always been possible, and emerge in the most various cultures (*AC* 4). There is no smooth path of modernity that we can trust to foster human improvement. If modern thinkers are right about anything, it is the fact of change, but they perversely construe change as intelligible development and sequence, covering over its radical, disruptive dimension. This recalls Nietzsche's other discussions of the importance of seizing the right time, or *kairos*. Moderns expect time ("development" or "progress") to present them with useful results. The disabused and vigilant know that opportunity must be grasped in the instant and that the "happy chances" must be treasured. Of course Nietzsche would also have us realize that opportunities will arise for us to take more concerted action; in the right circumstances, we can prepare grounds for the breeding and nurturing of "higher types," but "progress" will not do this for us.

So Nietzsche begins *The Antichrist* by dismissing the illusions of modernity, more sharply and concisely than he had in his *Unmodern Observations*. These illusions are not merely the mistaken and pernicious ideas that contemporaries happen to entertain but modernity's very philosophy of time. The book's last pages reject the entire Christian calendar and time scheme, whose most recent variations (as we shall see) are modernity and world-history. After calling Christianity the one great curse of humankind, Nietzsche adds: "And one calculates time from the *dies nefastus* on which this fatality arose—from the first day of Christianity! *Why not rather from its last?—From today?* Transvaluation of all values!" (*AC* 62). Let us disrupt the conventional reckoning of time, splitting it in two symbolically, as well as practically. This declaration is followed immediately by the "Decree against Christianity," which is dated as "Promulgated on the day of healing [*Heils*], the first day of the year One (—on September 30, 1888 in the false time reckoning)." In the Roman calendar, a *dies nefastus* was an unlucky or unfavorable day, on which citizens were not to engage in legal or political activities. So Nietzsche, adapting this expression from the highly political Roman calendar, marks once more *The Antichrist*'s insistence that Christianity and its progeny world-history are temporal-political missteps. The Romans, as Nietzsche knew well, did not regard their calendar as inviolable, and it changed several times before finally being replaced by the

Christian one. If the Romans could begin their calendar with the founding of the city, Nietzsche and his double Antichrist could begin a new one with their manifesto and decree.

"THE TRUE NAME OF THE ANTICHRIST."

If secularization does not really take place, so Nietzsche seems to reason, the path to a renewed vision of the human-earth could lie in deploying the conceptual resources and historical traditions of Christianity against it. I take this strategy to be an essential component of "a philosophy of the Antichrist." Now, let us expand Nietzsche's Antichrist dossier, recalling first a significant passage where he alludes to the Biblical apocalypse, and then considering his more explicit uses of the term itself.

For Nietzsche, both world-history and Christianity are hostile to the earth, the site of embodied, plural, mobile human life. This is glaringly obvious in the case of Christianity and only a little less so in that of world-history, which models itself on the Christian teleology that reaches its most extreme expression in Revelation. Christianity began by despising the earth. The problem of Jesus, who died too young, was that he did not learn to live and dwell on the earth (*Z* I.21). Yet early Christianity, when its initial apocalyptic expectations were disappointed, slipped within a century or two into making peace not with the earth but with the world, through the institutions of church and state. Both are described as lying and bellowing beasts in Zarathustra's speech "On Great Events," which in contrast praises the earth's heart of gold.

Given the constant Biblical resonances and parodies in *Zarathustra*, the chapter titled "The Seven Seals" needs to be read (as scholars have long noted) as alluding to the last book of the canonical New Testament. Zarathustra, who is several times associated with both Dionysus and Antichrist, is specifically linked with a central mystery of the Christian Revelation in "The Seven Seals," its title echoing a theme of that text. The Biblical seven seals are seals of a book, both hiding and disclosing—to the enlightened interpreter—the mysteries of first and last things. So Nietzsche too gives us a book for all and none, but in order to set up a counterscenario in which earth persists. If Christianity is a religion of the book, then the philosophy of the Antichrist has its own book, in which only earth is sacred. Specifically, Revelation depicts the end of the world and its worldly kingdoms (Rome, allegorized as Babylon). In the usual readings of the book, the Antichrist reigns for some time over the earth. After trials and tribulations, marked by war, earthquake, pestilence, and death, Christ will reign on earth with the saints until all is absorbed into God's eternal and heavenly kingdom.[12] Then earth disappears, once and for all. In Nietzsche's "Seven Seals," Zarathustra dies on and for the earth, knowing that he and earth will return.[13] While the Biblical end-times involve a series of disasters, Zarathustra's end is an ecstatic communion with the earth, in a flight touching sea and sky. In praising "free death," a self-chosen, self-timed death at "the right time" (*kairos*), Zarathustra had said that his own death and that of his friends should "glow like a sunset around the earth." "Thus would I myself die, that you friends might love the earth more for my sake; and into earth will I turn again, that I might rest in her who bore me" (*Z* I.21).

This voluntary death, then, is for the sake of the earth and for the friends, children, and disciples who have pledged their loyalty to the earth. It also recalls the Zoroastrian practice of *dhakma* or sky-burial, which Nietzsche surely knew of. Corpses were arranged in a circle at the top of a tower open to the air, left to scavenging birds or effects of the elements. Eventually, remaining bones were incinerated, and the lime was returned to the earth, to "nourish the life of significant soil." The purpose of *dhakma*, a practice that might strike us now as ecologically exemplary, is to avoid polluting either earth (in the narrow sense of one element) or fire.[14] This is Nietzsche's answer to Revelation: death can be self-chosen and can constitute a glorification of the earth. Already in *Zarathustra* we can see some lineaments of a philosophy of the Antichrist.

The concluding aphorism of "Peoples and Fatherlands," which emphatically declares that this is "the century of the *multitude*!" and describes the glories and miseries of the century's higher *Menschen* who exhibit Europe's growing cultural hybridity, asks rhetorically whether any of them could have been capable of "a philosophy of the Antichrist" (*BGE* 256). This fateful appellation also appears in the same year (1886) in Nietzsche's late "Attempt at Self-Criticism," added to the second edition of *The Birth of Tragedy*; there he says he was seeking an anti-Christian "doctrine and counter-evaluation of life . . . What was it to be called? As a philologist and man of words I baptized it, not without a certain liberty—for who can know the true name of the Antichrist?—by the name of a Greek god: I called it Dionysus" (*BT*, Attempt).

This is one of three statements Nietzsche makes concerning the linguistic sources or perspectives he speaks from in naming himself as the Antichrist. Naming itself is a significant activity or process for Nietzsche. At the extremes are "the lordly right of giving names" on the one hand, and that which is unnamed or unnameable, as in the "unnameable virtue" of which Zarathustra speaks.[15] The other, and apparently the first self-identification or, strictly speaking, naming, of himself as the Antichrist, appears in his letter to Malwida von Meysenbug, where he says that the name is taken from the language of the church (*Kirchensprache*). Nietzsche's correspondence with von Meysenbug was confidential and sometimes intimately playful and humorous. Obviously there are degrees of irony in the letter that announces, just between friends, his new name. The third and last linguistic gloss on this name is that in *Ecce Homo*: "I am in Greek, and not only in Greek, the *Antichrist*" (*EH* Books 2). The Greek prefix *anti* has a range of related senses: over against, in opposition to, mutually opposed to, in return, equal to, corresponding. We could read Nietzsche's linguistic reminder as indicating that he is more than simply the enemy of Christ; if he were just that, then he could be understood as taking a merely reactive stance, of voicing his ressentiment. This charge, of course, has been leveled frequently. Nietzsche also feels the need to record that this is his name "not only in Greek." So, I suggest, he takes on himself the name's association with catastrophe in the quite specific sense of a total turn or reversal, a cataclysmic upheaval.

These linguistic glosses on the *name* of the Antichrist, by the way, should be sufficient to cast doubt on those readings that would construe Nietzsche's use of the term as generally meaning nothing but anti-Christian.[16] In his 1954 introduction to his translation of *The Antichrist*, Walter Kaufmann said that "the title is ambiguous" and could either refer to the apocalyptic figure or simply mean "anti-Christian."[17] This is true of the German word *Antichrist* by itself, although context often clarifies by speaking of *der Antichrist* (*the* singular individual) or *eine Antichrist* (*an* anti-Christian).[18] The title of Kaufmann's influential 1950 book *Nietzsche: Philosopher, Psychologist, Antichrist* apparently uses the term as a name rather than an adjective, yet there is scant discussion of what this name might mean. While Kaufmann does have much to say about Nietzsche's mixed evaluation of Jesus and his much more vituperative attack on Christianity, he apparently does not mention his remarks about "a philosophy of the Antichrist." Perhaps in the 1950s, Kaufmann was unaware of the final page of *The Antichrist*, which is signed with that name. The page, suppressed by Elisabeth Förster-Nietzsche and her editorial associates, was restored in the now-standard Colli-Montinari edition. While Kaufmann later responds to that edition's version of *Ecce Homo* in the fourth edition of his own book (1974), he takes no notice of this addition to the text.[19] (Not surprisingly, Kaufmann also minimizes the intellectual significance of Nietzsche's friendship with Overbeck, Nietzsche's theological interlocutor and "comrade in arms," acknowledging at most that he "may have called Nietzsche's attention to helpful passages in early writers.")[20] Kaufmann suggests that Nietzsche was using the title Antichrist for sensational effect. This is undoubtedly correct, but there is more substance to Nietzsche's use than mere sensationalism. Perhaps Kaufmann himself was seeking a certain sensational effect in adopting the word in his own title. In addition to the texts already cited, Nietzsche calls himself the Antichrist in *Ecce Homo* (*EH* Books 2) and in a letter to Malwida von Meysenbug (*KSB* 6.357; April 4, 1883), and speaks of the man of the future, "this Antichrist and anti-nihilist, this conqueror of God and nothingness" who must come one day (*GM* II.24). The texts from the published works are considered in the following.

Traditional discussions of the Biblical term "Antichrist," as Kaufmann notes, typically observe that it may apply either to a singular figure or to anyone who takes up an anti-Christian stance, perhaps simply by denying Jesus's divinity, even if he is respected as an exemplary moral and ethical teacher. We need not take these as exclusive alternatives in reading Nietzsche, but it is difficult to avoid reading *der Antichrist* other than as *the* Antichrist. Moreover, as a singular figure, this enemy of Christ can be thought of somewhat differently, given the dual meanings of the Greek *anti.* The Antichrist could be either Christ's enemy or one who takes Christ's place, being a false parallel or double to Christ (so there are discussions of whether the Antichrist must be a Jew and have other characteristics mirroring Jesus). In either of these nonexclusive conceptions, the Antichrist takes his place as a fearsome ruler of the earth. Let us mention a further complication. Might the Antichrist be the book with this title itself? Andreas Sommer, who favors this reading, observes that Christianity is a "religion of the book," and Nietzsche, we know, expected his writings to have great power, an expectation that became especially focused on *The Antichrist* as he completed it.[21] Perhaps "a philosophy of the Antichrist" is resonant with all of these possible meanings.

The Antichrist is initially associated with the earth by Biblical texts and legend, since this will (for a time) be the site of his rule. Nietzsche goes further by divulging his guess at the Antichrist's true name: Dionysus, the figure of intensely earthly life, earth's renewal, and eternal recurrence. Not to be forgotten when Nietzsche invokes Antichrist are the word's legendary and vernacular associations, a well-known storehouse of widely distributed apocalyptic images produced by artists like Dürer and Holbein, and countless attempts to identify the figure with historical agents (including a Nietzschean hero, Frederick II).[22] All of these contexts involve a search for the Antichrist's name or identity.[23] When Nietzsche inscribed this term, especially in *Der Antichrist*, he knew that he was sounding the depths of the most powerful religious and philosophical tradition of the West. In a well-informed recent scholarly survey of the Antichrist theme in legend and theology, Bernard McGinn describes the eighteenth and nineteenth centuries as the low point of general interest in the figure of evil.[24] We might suppose then that Nietzsche's title would be more of a curiosity than an outrage. However, the late nineteenth century in Europe was a time in which many feared cataclysmic change. Socialist and Marxist revolutionary groups were organizing, leading to fear and uncertainty among those with property. Anarchist assassinations spread terror. Rumors of dangerous, conspiratorial secret societies had been widespread for a century (Marx and Engels called this "a specter haunting Europe"). By adopting the name Antichrist, Nietzsche may have been attempting to play upon the sense of impending catastrophe. In the twentieth century and since there has been a resurgence of explicitly religious apocalyptic fears and expectations, along with more reality-based apprehensions about nuclear war, climate change, resource shortages, and overpopulation. The appetite for stories of global disaster or apocalyptic themes appears insatiable. In the form of books like the best-selling *Left Behind* series of sixteen novels supposedly based on the Bible, or the many films with similar background, we can say that the Antichrist legend is alive and well. Ostensibly nonreligious versions of the apocalypse abound. Postapocalyptic films like *The Road* or *Mad Max* show us life after the next great wars. In the *Terminator* series we are terrified by what grows out of our own war-making and reaches back into our time to extinguish any chance of human survival. Films of climate-based disaster, like *The Day After Tomorrow*, purport to show us the consequences of our undisciplined burning of carbon fuels and other resource abuse. Then there are the terrors of biological engineering, from the zombie craze to Margaret Atwood's trilogy (starting with *Oryx and Crake*) that shows our world devastated by genetic modifications. Still other end-of-the-world scenarios involve nonhuman causes: unprecedented earthquakes, solar supernovas, collisions with meteors, asteroids, or encounters with extra-terrestrials. Nietzsche's adopting the name for the more "classical" imagination of disaster seems to presciently speak to both Christian and nonreligious fears of the world's end.

Nietzsche was, of course, familiar with the theological concept of the Antichrist, so his announcement and parodic self-identification as Antichrist has a theologico-political meaning. The book titled *The Antichrist* could be called Nietzsche's *a*theologico-political treatise, an anti-messianic response to world history and political theology that challenges not only the calendrical measure of *chronos*, which reinforces the presumptions of both of these

Equally importantly, he points to the degradation of the Christian *kairos* as an opportunity for total transformation; it now becomes the founding event of a church and a Christian-European world order. For Nietzsche it is no longer anno domini, and there are many dawns, many *kairoi*, yet to come.

From the late 1870s on, Nietzsche was regularly reading current scholarship on Christian beginnings and history, from scholarly studies of Paul, recommended (and often supplied) by Overbeck, to the more popular, novelistic, sentimental, and orientalizing series of books by Ernest Renan, *History of the Origins of Christianity*.[25] By the later 1880s, shortly after signaling the possibility of a "philosophy of the Antichrist," he was supplementing these studies with explorations in what we might call comparative political theology. The texts include Julius Wellhausen's philologically acute studies of Judaism and Islam, frequently cited and quoted in Nietzsche's notebooks. Wellhausen unearthed the buried conflict between priest and warrior in the history of Israel; this played into the Judaism that began to be formulated in the exilic period when priests reedited sacred texts, justifying their own power after the nation lost its political sovereignty (*GM* I.7, *AC* 25–27). Another touchstone for Nietzsche's late study of comparative political theology was the Indian Law of Manu, presented in the idiosyncratic French version of Louis Jacolliot. (Jacolliot's Manu book is announced as one of three on the three "great lawgivers": Manu, Moses, Muhammed, although the series was never completed.) To talk of Christ and Antichrist with a "trans-European eye" was to be critically engaged in comparative political theology.

Scholars still argue about whether *The Antichrist* was meant to be read as the first part of a four-volume *Transvaluation of All Values* or as the whole of that work (Nietzsche said both at various times, and we should be skeptical in reading definitive intentions into the scribbling and postcards of his final prebreakdown days.)[26] Without trying to answer this question, I propose to read *The Antichrist*, along with some of Nietzsche's other late texts on Christianity and religion, as constituting an attempt to reckon once more with the form of temporality specific to world-history. Nietzsche does this by exposing how Christianity, which originally offered an alternative to both the political time of imperial Rome and the theologico-political time of Jewish law, came not merely to compromise with the state and legality, but to furnish an impetus and paradigm for constructing experience, memory, and time within the philosopheme known as world-history. In the course of *The Antichrist*, Nietzsche describes and evokes other times, notably Jesus's blissful state of pure presence and the rapidly accelerating pace of Christianity's decline. "Breaking history into two" would be a temporal caesura, a shock of the highest order.

NIETZSCHE'S OUTRAGEOUS TEXT: NAMES AND TIMES

I read *The Antichrist* as integral to Nietzsche's struggle to overcome the idea of world-history and restore hope and future to earth and humans. But why take this outrageous book seriously? *The Antichrist* has probably done more than any other single text to convince readers and critics that Nietzsche's final years of madness, decay, and dementia were the realization of the direction in which he had been heading all along (we withhold judgment on whether the definitive cause was syphilis, uninformed drug use by Nietzsche and subsequent administration by doctors, or a slowly growing brain tumor that had finally reached a tipping point).[27] Yet it is difficult to base this appraisal on the book's hyperbolical tone alone, since the writer leaves no doubt that he is determined to shock and outrage. As he says in the final section (preceding "the Antichrist's" edict against Christianity), "Wherever there are walls I shall inscribe this eternal accusation against Christianity upon them . . . I call it the one immortal blemish of mankind" (*AC* 62). Now, this passage concludes by marking what I take to be a major and rather neglected dimension of this self-consciously excessive book, the question of time, a politically inflected time: "And one calculates *time* from the *dies nefastus* on which this fatality arose—from the *first* day of Christianity! *Why not rather from the last? From today?*—Transvaluation of all values!" (*AC* 62).

Christianity has established a claim to the measurement and naming of time. Within this tradition, whose crucial temporal axis is supposed to be marked by Jesus's birth, another more indeterminate date looms: the appearance of Antichrist and the associated events ushering in the end of the world. The calculation of time, the measuring out of *chronos*, I have been suggesting, is one of Nietzsche's persistent themes. It contributes to making the earth small, furthers the regime of the last humans, and produces a general blindness to the unpredictable future and the fleeting but genuine opportunities offered by chance. Yet we might ask whether merely substituting one calendar for another could be more than an ironic gesture if the primacy of calculation still goes unquestioned.[28] Perhaps Nietzsche wants us to wonder about this.

THE BIRTH OF WORLD-HISTORY FROM THE SPIRIT OF THE CHURCH

Christ and Antichrist, then, are names having to do with time, more specifically with calculated time. If Christians begin their calendar with Jesus's birth (even if postdated four years), they have also frequently attempted calculations of the world's end, signaled by the Antichrist's coming. These calculations began in earnest with the Church Father Hippolytus around 210 CE, calculations Nietzsche would surely have discussed with his housemate Overbeck, who wrote his doctoral thesis on Hippolytus's treatise *On Christ and Antichrist.* Overbeck's Latin dissertation concerns an important early patristic text in the tradition of eschatological speculation that grew up around the enigmatic references to the personification of evil. In the early third century, Hippolytus, living in Rome but coming from a Greek cultural background, published his short treatise defining several features of the Antichrist taken up by many later writers.

Hippolytus's originality was in providing arguments that claimed to render intelligible the deferral and delay of the Second Coming and the end of time, an increasing worry in the second and third centuries. Very few nontheological philosophers have even heard of Hippolytus of Rome, but he has had an enormous if indirect influence on the course of philosophy, arguably enabling the movement of thought that eventually crystallizes as "world-history."[29] Hippolytus contributed to codifying Church and European political theology for hundreds of years, consolidated and further defined the Christian picture of Antichrist and apocalypse, and, since the 1851 publication of his long-lost *Refutation of All Heresies* shaped the understanding of early Greek philosophy (perhaps it is fitting that this great thinker of deferral should have such a deferred effect).[30] His Biblical exegeses center around the names Christ and Antichrist.

The development of theology in the changing Christian community of the Roman Empire became Overbeck's lifelong theme, already foreshadowed in his dissertation on Hippolytus. Early theology began to ask questions like these: How and when would this world cease? How, in the meantime, can we live in a world of misery and pain? Eschatological beliefs, textual exegesis, and political theology contribute to various answers. Hippolytus expands the ideas of Irenaeus, his more original teacher. He shows a stronger interest in exegesis of sacred texts as a method of argument and exposition. Drawing especially on the Hebrew prophets, including the visions of Daniel, and scattered Christian texts (there was not yet a codified New Testament), he claims that as Christ is the recapitulation (*anakephalaiosis*) of all good, so Antichrist is the recapitulation of evil. At the same time, Antichrist must mirror Christ by appearing in human form, being Jewish, sending out apostles, and building a temple in Jerusalem (corresponding with the temple of Christ's body). Hippolytus engages in ingenious interpretation of 666, the number of the Beast.

What are these names or titles—Christ and Antichrist—that occupied Hippolytus? In what sense are they names? As Agamben reminds us (and as many—even theologians—have forgotten!) "Christ" is simply the Greek translation of the Hebrew "messiah," as in the Septuagint. "Christ" is a Russellian definite description, not a name. That should direct us to look more carefully at Nietzsche's use of the title and "name" Antichrist. Nietzsche, the philologist, former student of theology, and Overbeck's friend, would have understood that *christos* is not a proper name, but a title or appellative. As Agamben points out, a general ignorance (or willful forgetting) of this fact has enabled a downplaying of messianic thought and an oblivion concerning messianic time in later Christianity.[31]

In addition to signaling an all-out assault on Christianity, the Antichrist title should be understood within a tradition we now, once more, call political theology. Despite the obvious attempt to scandalize here, there is a theological history behind Nietzsche's declaration "I am in Greek, and not only in Greek, the *Antichrist*" (*EH* Books 2). The remark's context may seem to frame it as one more throwaway witticism, but the "old philologist's" emphasis on language usually repays attention. The section in *Ecce Homo* that works up to this claim begins with the less outrageous assertion that Nietzsche has many sensitive readers, especially outside Germany. He says that he is well received by those sensitive to style and nuance—those able to listen, such as his *Doktorvater* Ritschl and the French critic Hippolyte Taine. Perhaps women listen more attentively than men. He is not like an ass or donkey, whose long ears betray his crudity. And only after such strong provocations to attentive listening does Nietzsche come to his philologically mediated identification as Antichrist: "I am the anti-ass (*Antiesel*) par excellence and so am a world-historical monster,—I am, in Greek, and not only in Greek, the *Antichrist*."

This linkage of monstrosity, world-history, and the Antichrist can be understood in terms of a Christian tradition of political theology whose effect (and, doubtless, its occasional intention) is to obscure the consciousness of messianic time basic to the earliest Christian community. From the standpoint of world-history, Nietzsche acknowledges, he is a monster—but the careful listener might say, so much for "so-called world-history." Let us explore this conjunction of world-historical monstrosity and the linguistically inflected claim of the Antichrist name. Nietzsche's mention of Greek and unspecified other languages should lead to the question of the meaning and significance of the title Christ even before we consider the Antichrist figure. Nietzsche indicates that questions of time are at stake by calling himself a "world-historical monster" in the same sentence. As we know, world-history is itself something of a monstrosity for Nietzsche. What are the senses of time associated with the "idiot" Jesus, Christ as constructed by Christianity, the Antichrist and apocalypse, and the witches' brew of world-history that comes from the attempt to combine them? And what are the political implications of these forms of temporality? Can we demythologize the idea that the Antichrist is lord of the earth?

Nietzsche advises us to think of the name Antichrist in several registers, including at least its philological, philosophical, and theological contexts. Nietzsche calls for the Antichrist toward the conclusion of the *Genealogy*'s essay on " 'Guilt,' 'Bad Conscience,' and Related Matters." In the long penultimate aphorism leading up to the writer's falling silent before the "younger" Zarathustra, he imagines "the redeeming human of the great love and contempt" who must come in a stronger time:

> This human of the future who will redeem us from the previous ideal as much as from that *which had to grow out of it*, from the great disgust, from the will to nothingness; this bell-stroke of noon and of the great decision, that makes the will free again, that gives back to the earth its goal and to humans their hope; this Antichrist and anti-nihilist; this conqueror of God and of nothingness—*he must one day come* . . . (*GM* II.24).

The emphasis on the singularity of this anticipated figure certainly suggests a distinct individual, not simply one among others who are also opponents of Christianity. It is not merely the "previous ideal"—Christian ideals of compassion, selflessness, or self-abnegation—which is to be destroyed but, emphatically "that *which had to grow out of it*," that is, I take it, Christian culture and institutions, in all their multiform transformations and disguises (the *Genealogy*'s third essay traces these types all the way down to the *Wissenschaftler* and the historian). The Antichrist "gives back to the earth its goal and to the human their hope." The Antichrist is not merely destructive, then, and in an important sense he is not completely original, for he *gives back* to earth and humans what had been lost or taken from them. "Giving back" suggests return, revolving back to an earlier condition, one of the core dimensions of "revolution." The implicit contrast, then, is between the Antichrist, who restores these, and Christ (or Christianity, which claimed his heritage), who is responsible for their loss or theft. Returning to *Zarathustra*'s crucial terms, *Erde* and *Mensch*, Nietzsche closes this essay in the *Genealogy* by reminding the reader (who has been advised to read the earlier book first) of the thematic relation binding earth and human. The idea of world-history, then, contrasts with that of the human-earth (*Menschen-Erde*): one forecloses the future and subsumes the earth within a closed system of meaning, while the other reopens occluded horizons.

The Antichrist's hyperbolic rhetoric alienates many readers, yet Nietzsche foregrounds his hyperbole in designating his alert readers as "the fewest," as Hyperboreans who dwell far beyond the north, and are like him in "seeing the wretched ephemeral chatter [*Zeitgschwätz*] of politics and national egoism *beneath* themselves" (*AC*, Foreword and 1). To see such chatter beneath one is not to ignore it altogether, but to see it in its appropriate position, from a higher perspective. This perspective brings together two important critiques. These are directed, first, at the foundations of *Zeitgschwätz*, the all-too-timely or *zeitmässig*, namely the philosophy and ideology of world-history on which it is parasitic (cf. *UO* III). The second critique focuses on Christian political theology, a set of doctrines and concepts embedded in institutional practice, which have the effect of obscuring the life and temporal experience of the earliest Christian community while lending support to the worldly powers of state and church. World-history and political theology mask the possibility of a politics of the earth by enabling and giving a framework to the distracting *Zeitgschwätz* of the *Zeitungen*.

TIME'S DELAYS: KATECHON AND WORLD-HISTORY

Who is the Antichrist in the discourse of the church (*Kirchensprache*)? The answer will help us understand how Nietzsche's project of a "philosophy of the Antichrist" could be a further twist in his thought concerning place and time: that is, the earth and the world, *kairos* and *chronos*. The specter of the Antichrist helps to open up Nietzsche's perspective on political time (not mere *Zeitgschwätz*). Let's approach this question by considering the meaning of the figure in Christian political theology. This tradition claims its origins in letters attributed to Paul, was formulated explicitly as early as Tertullian in the second century, and has been explored by a number of more recent thinkers and scholars, notably Schmitt, Kantorowicz, and Agamben.[32]

In Christian political theology, as contrasted with the potpourri of passionate anathematizations and passing hysterias about supposed great events or cataclysms, it is not so much a question of naming the Antichrist as of understanding his, her, or their—for there can be many Antichrists, and perhaps by the logic of the simulacrum there must be—relation to political power. Briefly, in that long tradition, the Antichrist can be understood as the evil one(s), both enemies of Christ and figures capable of imitating his attributes, whose appearance is restrained or held off by the state. In the second letter to the Thessalonians, Paul (or whoever writes in his name) warns his addressees against expecting an imminent arrival of the last days, the end of the world. He says without much explanation that there is a restraining force or *katechon* (*Aufhalter*) that delays the appearance of Antichrist(s):

> Let no man deceive you by any means: for that day shall not come, except there come a falling away first, and that man of sin be revealed, the son of perdition; Who opposeth and exalteth himself above all that is called God, or that is worshipped; so that he as God sitteth in the temple of God, shewing himself that he is God. Remember ye not, that, when I was yet with you, I told you these things? And now ye know what withholdeth [*katechon*] that he might be revealed in his time [*kairo*]. For the mystery of iniquity doth already work: only he who now letteth will let, until he be taken out of the way (II Thessalonians 2: 3–7).

This is the rather enigmatic text that Tertullian applied to the Roman Empire around 200 CE, roughly a century before its official Christianization under Constantine. The theme of waiting, containing, warding off, and other apotropaic strategies has more general implications, and can be understood in such contexts as the Cold War doctrine of containment, slowing climate change, resistance to reform by entrenched institutional interests, or the struggles of those with aging bodies to mitigate the processes of decay and degeneration.[33]

In his *Apology* Tertullian writes:

> There is also another and a greater necessity for our offering prayer in behalf of the emperors, nay, for the complete stability of the empire, and for Roman interests in general. For we know that a mighty shock impending over the whole earth—in fact, the very end of all things threatening dreadful woes—is only retarded by the continued existence of the Roman empire. We have no desire, then, to be overtaken by these dire events; and in praying that their coming may be delayed, we are lending our aid to Rome's duration.[34]

When the Roman Empire becomes Christian and, later, Europe is ruled by Christian kings, such katechontic teachings become increasingly significant, providing ideological support for royal and imperial authority. It is not surprising then that Schmitt, Kantorowicz, and Agamben find the doctrine of the restraining force of the *katechon* essential in the claims to legitimacy put forward by medieval and early modern European states. For Schmitt, the state legitimated in terms of this teaching is the emblem and central institution of the temporal regime that began with the coming of Christ and will end in the world's last days as foretold in Revelation:

> The Christian empire was not eternal. It always had its own end in view. Nevertheless, it was capable of being a historical power. The decisive historical concept of this continuity was that of the restrainer: *katechon*. "Empire" in this sense meant the historical power to restrain the appearance of the Antichrist and the end of the present eon; it was a power that withholds (*qui tenet*), as the Apostle Paul said in his Second Letter to the Thessalonians.[35]

Katechontic political theology involves a very specific sense of *temporality*, one that helps to constitute the Christian core of *Weltgeschichte*, which Nietzsche saw as the dominant form of his age's temporal thought. Katechontic (and world-historical) political time contrast with the temporal consciousness of earliest Christianity, the community of Jesus's followers, who believed that he had ushered in a new, messianic time. Paul writes: "But this I say, brethren, time contracted itself, the rest is that even those having wives may be as not [*hōs mē*] having, and those weeping as not weeping, and those rejoicing as not rejoicing, and those buying as not possessing, and those using the world as not using it up. For passing away is the figure of this world" (I Cor. 7:29–32).[36]

We are to undertake all activities (work, marriage, civic relations) "as not" these activities. Time will be transformed. Yet as Agamben says, "a messianic institution—or rather a messianic community that wants to present itself as an institution—faces a paradoxical task."[37] Institutions and communities are meant to endure. What then is an institution or community that expects an imminent end of all worldly institutions and communities? Agamben observes that both the Christian and Jewish traditions have entered an implicit agreement to see Paul as the founder of a new religion. In doing so, "the aim is to cancel out or at least mute Paul's Judaism, that is to say, to expunge it from its original messianic context."[38] Agamben cites Jacob Taubes's 1986 final lectures (and draws on his own vast erudition) to support his stringent rereading and reevaluation of Paul. However, Nietzsche (cited extensively by Taubes) was already aware of the discrepancy between earlier and later responses to the Jesus event.

FROM OVERBECK TO NIETZSCHE

Martin Heidegger's student Karl Löwith wrote an ambitious and illuminating narrative titled *From Hegel to Nietzsche: The Revolution in Nineteenth-Century Thought*. The book artfully combines aspects of a sequential history with a structure based on an array of philosophical topics. The last section is devoted to "The Problem of Christianity." Running his analysis again, more or less chronologically, through a number of major figures, he concludes the section and so the entire book not with Nietzsche, but with "Overbeck's Analysis of Primitive and Passing Christianity." Of course Overbeck, who brought Nietzsche back to Basel after his collapse in Turin, had a productive life that continued for about fifteen years more. Löwith sums up his contribution as having achieved a clear understanding of why and how the Christianity of the bourgeois world came to an end. Yet since Overbeck's researches demonstrated "the abyss separating us from Christianity," Löwith ends his book with a qualification and a rhetorical question: "This does not mean that a faith which once conquered the world perishes with its last secular manifestations. For how should the Christian pilgrimage *in hoc saeculo* ever become homeless in the land where it has never been at home?"[39] The expression *in hoc saeculo* is apparently a reference to Ephesians 1:21, where Paul writes (in the King James version), describing the power given Christ by God: "Far above all principality, and power, and might, and dominion, and every name that is named, not only in this world, but also in that which is to come."

The Greek term that the King James translates with "world" (and Luther with "*Welt*") is *aion*, also and more recently often translated as "age." The difference may not be very significant, since the world in Pauline perspective is always only that of an age or era that will give way to "that which is to come." Löwith, I assume, is playing on the relation between the *saeculum* and the secular or *weltlich*. He endorses Overbeck's view that primitive Christianity was rigorously unworldly. At the same time, we can read something different in Paul's letter, from which Löwith draws this parting question. Yes, Christ's kingdom is not of this world or age, Ephesians implies, but it stands not so much or only *outside* them, but *above* them. It was openings like this that helped to make Christianity into the "secular" power that Löwith assumed (in 1939) it no longer was. Given the resurgence of religion and political theology, it should be worth exploring just how Nietzsche and Overbeck engaged with the problematic of Christianity and worldly power.

Nietzsche's *Bildung* prepared him to understand the striking archaeological gap between the early community and the church that eventually emerged.

We will follow that understanding by placing him alongside his scholarly shadow Overbeck, whose work showed the way to that understanding in a voice that was as cautiously reserved and balanced as Nietzsche's was deliberately flamboyant and excessive. As we know, Nietzsche was brought up in a Lutheran pastoral family with ministers on both father's and mother's sides. He began university studies in theology before switching to classical philology. By 1865 he was having vigorous arguments with his sister Elizabeth, fueled by his reading of Strauss's *Life of Jesus*. It was the same year that he began to wrestle with Hegel's *Weltgeschichte*. In any case, he certainly discussed early Christianity personally and in correspondence with Overbeck, for whom this contrast of earlier and later formed the basis of his lifelong investigations of Church history.[40]

Overbeck was a penetrating student of the early theology and practice of the church, which he saw as attempting to make sense of the time opened up by Jesus's failure to reappear by constructing a narrative of historical time, building on such scanty and enigmatic Biblical passages as those on Antichrist.[41] The 1870s and 80s were marked by a flurry of critical and philological discoveries about early Christianity, its relations with Roman power, and the formation of the Biblical canon. Overbeck was a major player in this research activity. By 1886 Overbeck had informed Nietzsche that recent scholarship had identified Revelation as the rewriting of a Jewish apocalyptic text.[42]

In lectures on Gnosticism, Overbeck points out that in 150, a canon was unknown to those now considered mainstream Christian writers and Church Fathers, but by 200 it was more or less in place, although the process of its establishment was murky. Overbeck argues that it was the contest with Marcion's Gnosticism and its canon—which excluded the Hebrew scriptures—that led to the formation of what has since been recognized as the canon.[43]

In the struggle with Gnosticism, the Church developed a theory of time (as real and continuous rather than punctuated by absolutely abrupt revelations), a political structure (to combat heresy and order lives in the world), and an accommodation with the state (rather than dismissing it as merely an illusion of the fallen world).[44] The logic of theology as well as the strategy for suppressing Gnosticism pointed the Church in the direction of organization, doctrine, and practice that acknowledged history as it made its peace with the state.

As Nietzsche, the philologist, points out, presuming to name the Antichrist is a risky undertaking. Historically, the list of candidates has been extensive; a quick look at the Internet shows that almost any significant political figure has been held to deserve the title, beginning with Nero, infamous persecutor of Christians (and philosophers). More recently, some pointed out that Ronald

Wilson Reagan, each of whose names has 6 letters, displayed 666, the number of the Beast of the *Apocalypse*. During the thirteenth-century struggle between Emperor Frederick II (one of Nietzsche's heroes) and the Pope, both sides argued that the opponent was Antichrist.

Jacob Taubes says that "Overbeck's voice was heard much more clearly from the grave than it was by his contemporaries."[45] The scope and depth of his research and his reevaluation of Christianity and its theology have begun to become evident only in recent decades. Before a critical edition of Overbeck's works began to appear in the 1990s, he was known mainly through a few very detailed philological and historical studies of early Christian texts and authors, and as "Nietzsche's friend." A few thinkers, notably Karl Barth, were deeply affected by the posthumous (and editorially problematic) compilation of some of his notes published in the 1920s. Despite Karl Löwith devoting the final, summary chapter of his magisterial *From Hegel to Nietzsche* to Overbeck, little has been done to clarify his intellectual relationship with his more famous friend. Yet surely it was Nietzsche, if anyone, who would have been familiar with Overbeck's program before the latter's posthumous work began to appear. In 1873 Overbeck published his iconoclastic *How Christian Is Our Present-Day Theology?*[46] He argued that Christian theology, in accommodating state and history—notably from Paul to Constantine—was radically incompatible with the original Christian community's temporal consciousness. The latter involved rigorous rejection of the life and values of the "world," coupled with firm faith that its members were living in the last days before Christ's Second Coming, to be ushered in by catastrophes and tribulations forecast in apocalyptic texts. For a community living in expectation of the imminent end, accepting Jesus's counsel to take no thought for the morrow, to engage with the world "as not" in the world, theology would have been meaningless or superfluous. Nietzsche's portrait of Jesus is even more radical; Jesus lives in a pure present without any expectation of the last days, which was an addition by Paul and the early Church. John's first letter counsels, "Do not love the world [*kosmon*] or anything in the world" (1 John 2:15), yet Christianity invents world-history as part of the complex process of its accommodation to the world. Overbeck thought that this must eventually lead to a rejection of Christianity; as he wrote to Heinrich Treitschke: "If there is one point where one realizes that Christianity has twisted humanity, it is to be found in all the connections that Christianity has entered into with politics, just as I do not doubt at all that here is the point at which it will one day succumb to a general contempt."[47]

For Overbeck, the most notable survival of the spirit of primitive Christianity's rejection of the world and its time was the monastic tradition in its most ascetic and unworldly varieties. Surveying the modern epoch, he (like Nietzsche) singles out Pascal as a unique Christian who managed to recreate something like the antiworldly spirit of the first generations. For both, Pascal is the exception who demonstrates the degradation that has all but obliterated the earlier movement. Parallel to Nietzsche's attack on Strauss's "cultural philistinism" and Hartmann's *Weltprozess*, Overbeck saw contemporary liberal German theologians as shameless apologists for Bismarck's *Reich*. Later, thinking of his arch-foe Adolf Harnack, who supported the *Reich*, Overbeck wickedly suggested a parallel with Eusebius, probably Christianity's first political theologian and certainly its most effective one, who provided the chief formulations for Constantine's "donation" of the Roman Empire to the Church. Both, he said, were "*friseurs* of the Emperor's theological periwig."[48]

In Overbeck's narrative, when apocalypse was delayed past the generation who would see the coming of the kingdom in their lifetime, as Jesus preached, explanations seemed in order. Followers of Jesus, as Overbeck insistently repeats, were at first deeply convinced of the misery of earthly life, and they aspired, as in the Acts of the Apostles, to live in loving communion, their compassion for others rooted in freedom from "the world." But as time passed, the Church increasingly made worldly concessions, eventually appropriating concepts of Greek and Roman philosophy to rationalize its position in the continuing fallen world. This struck Overbeck as an existentially bizarre transformation of early Christians' sense that they were living in messianic time, that the *kairos* had arrived, that they were living in the time of the end, that the end-times were near. In his commentary on Paul, Agamben explains that this was conceived as the time of the end, as distinguished from the end of time. Now, in Paul's terms, time had been contracted; *kairos* on this understanding is a contraction of *chronos*. Yet after the initial generation or so of the Christian community had died out, and at least through most of the second century, the meaning of Christianity "itself" was radically up for grabs, as multiple versions and "heresies" (as they were called retrospectively), including Gnosticism and various Judaizing and anti-Judaizing sects, claimed legitimacy and authority. Overbeck's research concerned the formation of the Biblical canon, which went hand in hand with the consolidation of the Church, reaching a point of crystallization in the theology of Eusebius and Constantine's "donation" of the Roman Empire. Overbeck said that the upshot of this ferment, given form through fixing the canon and making Christianity the imperial Roman religion, was a Christianity that was nothing more than "the last phosphorescent glow of the decomposing ancients."[49]

There is an intriguing parallel between two texts on the Antichrist. Overbeck began his career with a dissertation on Hippolytus's treatise on the Antichrist, as Nietzsche completed his with a declaration of war and a curse on Christianity, which he titled and signed with that name.[50] Overbeck, already distanced from traditional Christianity and feeling incapable of preaching and pastoral work, opted for the scholarly refuge of theology and teaching. As he wrote Treitschke, he chose Hippolytus as a subject because he feared that dealing directly with canonical New Testament writings could have involved an aspiring professor in conflicts for which he was not yet ready. At the opposite pole, Nietzsche burned his academic bridges early with his *Birth of Tragedy*, and finally tempted the censors, flaunting the title *Antichrist* to signal that he was breaking history in two.

Even if young Overbeck was in part motivated by prudential concerns about his scholarly career, his choice of topic turned out to be significant for his later work, which focused on the contrast between early Christianity's awareness of messianic and eschatological time on the one hand and the later Christian accommodation to the world on the other. From Overbeck's perspective, early Christianity was fixated above all on the misery, evil, and corruption of the world; it anticipated a rapid transformation through Jesus's Second Coming. When this event was delayed, well past any living memory of Jesus's times, accommodation with the world became necessary. Part of this accommodation was theology itself. The very attempt to pose and answer questions about religious matters inevitably involved admitting some of the world's language and traditions of reasoning. Overbeck never stopped pointing out the ways in which Christianity was forced to use pagan learning, and so, he sardonically observed, it became the vehicle through which that culture survived. In Nietzsche's more succinct words, Christianity became "Platonism for the people."[51]

From the beginning of his work, Overbeck was inspired by a view that Nietzsche articulated much more dramatically in his later writing. Nietzsche's friend saw Christianity as necessarily containing the seeds of its own destruction. The title of Overbeck's posthumously edited "book" *Christianty and Culture* (a selection of his writings) could be taken as posing the exclusive alternative, Christianity *or* culture? His research focused on Church history, seen as the process by which Christianity accommodated itself to the world. Yet terms must be clarified, for in the early centuries there was not one Christianity but a multiplicity of sects and groupings with competing social, sexual, and political practices, beliefs, and texts. The "world" was, by and large, the Roman Empire. Only with the Council of Nicaea and the donation of Constantine did the Church achieve a tenuous stability in doctrine and structure. This complex history is the context for Overbeck's typical questions: How was the authority of the Church established? How were the canonical texts canonized?

Perhaps Hippolytus's greatest innovation was his redating of the final events to about five hundred years in the future. This opens up a time of waiting, expectation, and the possibility of accommodation to the world in the time that remains. From his reading of Daniel he deduced, for example, that there were still ten democracies and ten kingdoms yet to come, before the end of the world.[52] As I have been claiming, it is something like this time of waiting (*chronos*), which for Overbeck is both symptom and enabler of accommodation to worldly culture, and for Nietzsche a time that threatens to extinguish a vigilant watch for the opportunities (*kairoi*) offered by chance (*BGE* 274). Part IV of *Zarathustra* can be read as enacting a parody of the waiting that became basic to the Christian tradition. At the beginning of this section, Zarathustra explains his receptivity to the humans who may come to him in his mountain home:

> I still await the sign that it is time for my descent . . . not impatient, nor patient, but rather as one who has unlearned even patience—because he no longer "endureth" [*duldet*] . . . for my fate is leaving me time . . . I am grateful to it, my eternal fate, for not rushing and pressing me, but leaving me time for jesting and wicked tricks . . . (*Z* IV.1).[53]

The rest of Part IV is given over to Zarathustra's "jesting and wicked tricks" with the higher men who come seeking him. This is not the resigned Christian endurance of life in a vale of tears, but a time that has opened and dilated for the pleasures of play. Waiting has been transformed into playfulness that enjoys rather than merely enduring.

Even after their common residency in Basel, and after Nietzsche had taken leave from the university there, he continued his reading of Overbeck's work and related research. Some of the signs are evident in *Dawn*, which contains an extended discussion of Paul entitled "The first Christian" (*D* 68), as well as many observations on Christianity inspired both by Overbeck and other contemporary scholars. Nietzsche worked on *Dawn* in 1880. The work's composition was accompanied by a flurry of correspondence with Overbeck concerning the latter's research and other recent work on Paul and early Christianity. In June 1880 Nietzsche writes to Overbeck that he has reread

How Christian Is Our Present-Day Theology? and has understood much more than on his earlier reading, a reading that had already led them in 1873 to call their two-pronged assault on D. F. Strauss and liberal Protestant theology a *Waffengenossenheit*, or comradeship in arms. In July Overbeck sends Nietzsche his relatively brief new book *On the History of the Canon*, along with others' books on Paul's anthropology and Justin Martyr. He writes that the studies in this book have the value for him of "preliminary works on the emergence (*Entstehung*) of the earliest Christian literature, and are important for that of the canon and its clear understanding."[54]

Already, before *On the History of the Canon*, Overbeck had been engaged in an acutely political reading of early Christian literature. In 1870 he published an introduction to The Acts of the Apostles, arguing that the text was written considerably later than the activities recounted there, and that the narrative had been shaped to suggest the compatibility of Christianity and the Roman Empire. I quote at some length to give the flavor of the analysis. Overbeck points to

> the political side of the Acts—its obvious striving to procure for the Christian cause the favor of the state authorities of Rome by the consistent representation of the good terms on which the personages of the Apostolic period, particularly Paul, stood with the Roman state and its officers . . .
>
> Especially does the trial of Paul give the Roman officials the opportunity of showing the favorable opinion they have of him; and, shielded by the Roman laws, he is enabled, though still a prisoner, joyfully for a considerable time to fulfill in Rome his duties as an Apostle . . . Nay, the long detention of the Apostle is in part explained simply by forgetfulness of duty on the part of certain officials . . . In this account, to which the experience of Paul can hardly have corresponded, we cannot fail to recognize the design to avert political suspicions from Christianity, and such an account, in the form presented in the *Acts*, cannot have been intended for any other address than the Gentiles outside the Church.[55]

If such early texts are suspect, their canonization is even more so. Although the explicit subject of the longer of the two essays constituting *On the History of the Canon* is the question of what can and cannot be known about the date and authorship of the Letter to the Hebrews attributed to Paul, it begins with a rather far-reaching declaration that might equally well have been written by Nietzsche: "It is part of the essence of all canonization to render its objects unknowable, and so it can be said of all of the writings in our New

Testament, that in the moment of their canonization they have ceased to be understood."[56] Nietzsche was given to stronger formulations: "Paul understood the need for the lie, for 'faith'; the Church subsequently understood Paul" (*AC* 47).

Overbeck's book on the canon continues in the spirit of his analysis of Acts, demonstrating the philological acuity that Nietzsche so respected and admired. The motto of philology in Nietzsche's and Overbeck's tradition might be "always more than one language" (paraphrasing Derrida). Philology is critical and diacritical, always marking differences in and within authors, texts, traditions, and interpretations. Overbeck's critical hermeneutical sensibility was aroused by discrepancies between Hebrews and Paul's other letters.[57] Especially obvious (Overbeck is not the first to note this) is that the letter is composed in much more fluent Greek than the other letters, whose Greek is rather awkward (as Nietzsche wrote, "It was subtle of God to learn Greek when he wished to become an author—and not to learn it better" [*BGE* 121]).[58] To be canonical it was required that a text's author be one of Jesus's apostles. Overbeck concludes that the text was attributed to Paul so that it could be pronounced canonical. The desire to canonize came, he argues, because the Church found its content important and useful in securing its doctrines in opposition to Gnostic tendencies like Marcion's. It could also appeal to Jews who might "come to Jesus" if they felt that, far from surrendering their tradition, they could see this as its fulfillment. Nietzsche had read Overbeck's critical analyses. He recapitulates such inquiries from his own "physiological" perspective when he writes that "one is not philologist and physician without also being at the same time *anti-Christ(ian)* [*Antichrist*]. For as philologist one sees behind the 'sacred books,' as physician behind the physiological depravity of the typical Christian. The physician says 'incurable,' the philologist 'fraud'" . . . (*AC* 47).[59]

TIMES AND TEMPORALITIES OF CHRIST AND ANTICHRIST

Overbeck's research and Nietzsche's judgments of Christianity coincide in emphasizing a radical break between the temporal experience and expectations of the early community and the church that slowly took form in the next few centuries. Katechontic time, the time of waiting and deferral, has obvious political implications. With church and state established and coordinated, Christianity finds the basic lines of temporal life defined: the state, with the church's endorsement, resists the coming of Antichrist. History can now be plotted in terms of a story of deferred redemption, with regular payments on the debt in the form of confession and penance. Yet the time of the *katechon* is only one mode of medieval Christian thought about time and history. At least since Bergson and Heidegger, philosophy has been aware of distinctions like those between chronological time and the lived experience of *dureé*, or between an authentic grasping of future possibilities and self-deceiving acceptance of the "objective" time of *das Man*. Nietzsche's writings abound in accounts of various phenomenological modes of temporal experience. He frequently reminds us that his many styles of writing are intended to convey varying tempos of experience. What needs emphasis is that this sense of temporal plurality is not restricted to purely individual experience. There are social, collective ways of marking and structuring time, not only by measurement and punctuations (for example, by holidays and intercalary days) but in terms of speeding up or delaying, contracting or dilating.

Historians of philosophical and religious thought, like Löwith and Taubes, see the tendency toward condensing the many varieties of temporality as stemming from the apocalyptic and eschatological commentaries and speculations of Joachim of Fiore, who elaborated a three- stage conception of human history consisting of the epochs of God the Father, the Son, and the Holy Spirit. For Joachim, the age of the Holy Spirit was to begin in 1200. Such critical studies lead Agamben to sharply distinguish messianic time (the time of the end) from eschatological time (the end of time).[60] Messianic time is experience free of obsessive hope, regret, and nostalgia; it lives in an expanded present, not in waiting or expectation of a future state. Eschatological thought expects, awaits, and frequently attempts to predict the ultimate end. Messianic time is apolitical. Eschatological time requires an interim politics adapted to the specific character of the destined end. For Nietzsche, it was Paul, the evangelist and community organizer, who laid the foundations for eschatological time and its politics.

Nietzsche hyperbolically restates Overbeck's thesis about the later oblivion of messianic time in Christianity when he writes, "The word 'Christianity' is already a misunderstanding—in reality there was only one Christian and he died on the cross." In explicating this famous bon mot, Nietzsche distinguishes between living "practice" and mere "belief," just as Overbeck distinguishes the first community around Jesus from the Church and its doctrines:

> The "Evangel" died on the cross. What was called "Evangel" from this moment onwards was already the opposite of what *he* had lived: "*bad* tidings," a dysangel. It is false to the point of absurdity to see in a "belief," perchance the

> belief in redemption through Christ, the distinguishing characteristic of the Christian: only Christian *practice*, a life such as he who died on the cross *lived*, is Christian . . . (*AC* 39).

The "belief" in question here is a belief about what is to come, at the end of one's life or the end of time. The Stoics criticized time experienced as mere waiting and deferral, a critique that can be applied to the Church's conception of katechontic time as that during which governments ward off the coming of the Antichrist. Stoics aimed at eradicating hope and fear, both of which blind us to the lived experience of the moment and the readiness for real opportunity or *kairos* when it arises.[61] As Seneca writes: "Two things must be cut short: the fear of the future and the memory of past discomfort; the one does not concern me anymore, and the other does not concern me yet."[62] Jesus, as Nietzsche understands him, was free from time-as-deferral but lacking a sense of *kairos*; this reading supposes that the idea of the imminent arrival of the kingdom, understood as total transformation or end of the world, is already an invention of Paul and the early Church. Nietzsche engages in an abductive inference, leading him to his idea of Jesus as a Dostoyevskian "idiot," living blissfully in the moment, announcing the kingdom of God and time's accomplished fulfillment, totally indifferent to institutional structures as he declares that we should "take no thought for the morrow." This blank canvas allows the inscription of history's palimpsest and is very close to Overbeck's picture of early Christian community psychology.[63] In Nietzsche's counternarrative, Christianity was Paul's invention, positing Jesus's resurrection as history's turning point and the basis for a new, potentially universal faith community of both Jew and gentile. Nietzsche's parody of Paul (*The Antichrist*) offers another way of splitting history in two.

Nietzsche's deliberate caricature of church and theology coincides with the critique that Overbeck pronounces in more restrained but no less cutting fashion. The typical Christian theologian (in the spirit of Hippolytus) constructs a story about Jesus and the Church that accommodates the persistence of mundane time when the *eschaton* fails to arrive. The theologian makes his peace with the "world," which the earliest Christian community prided itself on dismissing. The theologian *must* be concerned to *sublimate* the various forms of what Walter Benjamin called *Jetztzeit* that characterized primitive Christianity, that amalgam of Gnostic salvation, Stoic *kairos*, and Jewish messianic time. Eventually theology leads to the invention of world-history. Some of Nietzsche's most vicious remarks about the theologian as a type emphasize the theological core of German idealism. Kant was so well received by the

Germans, Nietzsche says, because with his thought "a secret path to the old ideal stood revealed" (*AC* 10; cf. 8–11). That idealism culminates in Hegelian world-history, which Nietzsche had been in the business of bracketing, exposing, and undermining since at least the *Birth of Tragedy*.

At the heart of Nietzsche's *Antichrist*, then, is the rejection of the theological time of waiting, deferral, and gradual culmination. Christian time eventually became world-history, whose core is the state, "God's march through the world," as Hegel said. From this perspective, the final page, the "Decree against Christianity"—signed "the Antichrist"—symbolically announces a new calendar and new experience of time. If time was measured previously from Jesus's birth, a new beginning is possible after the demolition of Christian world-history.

The last paragraph of the main text of *The Antichrist* promises a writing on the walls and a revision of the calendar based on Nietzsche's declaration of war on Christianity: "Wherever there are walls I shall inscribe this eternal accusation against Christianity upon them—I can write in letters which make even the blind see . . . And one calculates *time* from the *dies nefastus* on which this fatality arose—from the *first* day of Christianity! *Why not rather from its last?—From today?*—Revaluation of all values! " (*AC* 62).

In *Difference and Repetition*, Deleuze writes of the profound structure that underlies this phantasm of the splitting of time. He contrasts the Cartesian "cogito" and the Kantian "I think." The cogito determines the "I" as a thinking being whose identity is dependent upon God. Kant saw that only the bare form of temporality was implied by the "I think," and so consequently "time signifies a fault or a fracture in the *I* and a passivity in the self."[64] Rational psychology must go the way of rational theology. Hölderlin and Nietzsche are the true heirs of Kant, rather than Fichte and Hegel. The time is out of joint and necessarily marked by a caesura. This "caesura, of whatever kind, must be determined in the image of a unique and tremendous event, an act which is adequate to time as a whole."[65] So Nietzsche's interruption of time, whether in the thought of eternal recurrence or in the introduction of a new calendar, is to be understood as a "symbolic image" of time out of joint.

The Antichrist's time is war-time, the war against Christianity and its "false reckoning of time." It is the war that follows in the wake of God's death. War-time cannot be planned, plotted, or predicted—or rather, these actions can all be attempted in full knowledge of their impossibility. In the parable of the madman, time is out of joint because the true news of God's death, like the light from distant stars, is still on its way to the traders in the marketplace. What they do not see, in their all-too-easy, all-too-human atheism, is that their

Wall Street Journal is always already out of date, and that the market is deeply complicit with the more radical risks of war. In war we become other, reduced to our position in the ranks or reidentified with a nom de guerre. Nietzsche's war-time is the time in which "I is another," in which the Antichrist repeats the tragic age of the Greeks in its post-Christian and post-Cartesian difference.

The declaration of war, the "Decree against Christianity," removed by Nietzsche's sister and early editors, was finally restored by Colli and Montinari.[66] Its title and form no doubt owe much to his studies in comparative political theology, from the interests he shared with Overbeck to his fascination with Jacolliot's book *The Religious Lawgivers* and its eccentric translation of the Law of Manu along with quasi-Darwinian commentary. The "Decree," meant to form the last page of *The Antichrist*, could be the text of a public poster, defining the law of an occupying power—or as the parody of such an act, emphasizing the political thrust of the text. It declares that priests should be imprisoned, that "every participation in a religious service is an attack on public morality," that Protestantism is worse than Catholicism and liberal Protestantism its worst variety, that all preaching of chastity or denigration of sexuality are to be condemned, that eating with a priest is forbidden, and that ostensibly laudatory Biblical words like "God" and "redeemer" should be understood as insults. It is impossible to determine what in this decree is parodic and what is meant in total seriousness. Nietzsche, who constantly tells his readers that he expects the most subtle, philologically attuned ear for his writings, offers something in comic book or graffito style, a set of directives as shockingly simplified, if not more so, than the instructions and videotapes of many recent terrorists.

The dateline immediately following the title of the "Decree" reads, "Proclaimed on the first day of the year one (—on September 30, 1888 of the false time scheme)." Nietzsche's career could be said to culminate in his failed project to recalibrate earthly time. He cannot avoid citing what he calls the old, false system of reckoning time. It should be day one, year one, of a time that has been newly divided in two—a new common era. Nietzsche's day one is put forward as an affirmative date, yet to the extent that it recognizes a prior time with which it marks a contrast, the question arises whether it can indeed be completely affirmative. In declaring war against Christianity, he must mark time with a new calendar that is not indebted to the very system of values he is combating. But like Christianity and Islam, Nietzsche's calendar splits the history of humanity into two, unlike the Jewish calendar, which begins with the creation of the world and so has nothing anterior to its basic date. Nietzsche's new way of reckoning time is tied up with wars and battles.[67] But then

Christianity, as Nietzsche describes it very explicitly in the concluding pages of *The Antichrist*, has always been at war. It has always been political, at least since the Church transformed Jesus into Christ. Whether surreptitiously conducting its vampiric raids on Rome, its Crusades against Islam, or its cultural wars—from Luther's attack on the Renaissance to Bismarck's *Kulturkampf* on behalf of the *Reich*—it is indeed a war-machine masked by its professions of peace.

The Antichrist began by taking its (or his) distance from *Zeitgeschwätz*, or the chatter of the media. It ends with a rapid-fire, scathing review of Christianity as an embodied historical and political force, announcing a new way of reckoning time. Media chatter takes place unreflectively within the world; at its most ambitious, it aims at positioning itself within its ultimate context of world-history. The final condemnation of Christianity and the curse upon it portray it as an infamy on the earth. The "Decree" goes so far as to declare that churches must be regarded as pollutions or miasmas of the earth. To shift the focus from church and state to the earth they pretend to subsume and rule involves at least a sketch of what a "great politics of the earth" might be, and what kinds of things might be termed "great events." Ever since Nietzsche introduced that term in his last *Unmodern Observation*, describing what he then took to be the tasks of Alexander and Wagner, it had a scope that surpassed the "world-history" on whose discovery the nineteenth century so prided itself. Insofar as *The Antichrist* constitutes all or part of the project of transvaluing values, it raises the question of the earth's future in a way that neither the op-eds of the journalists nor the world-historical speculations of the Hegelians could. At the very edge and distant horizon of the Christian and world-historical traditions, forgotten, ridiculed, or marginalized by their modern heirs, the paleonym (or older name) of Antichrist was a way of thinking about earth's future on a larger scale, however confusedly. Perhaps it is not surprising, then, that apocalyptic thought tends to resurge in times that fearfully anticipate climate change, transnational religious war, shortages of basic resources, new waves of migration and nomadism, as well as rapid and unpredictable technological transformations that penetrate deeply into all social relations. Unlike those who scour the Biblical texts for literal, all- too-literal, news of the last days, Nietzsche's donning of the Antichrist mask can be read most fruitfully as a way of asking now, at this date—is it early or late?—how we might begin to think earth's times in ways that do not foreclose being open to great events and great politics.

NOTES

CHAPTER ONE

1. See key to references.

2. In a vita written for Georg Brandes, Nietzsche marked his own genealogy by associating it with the decisive battle of Lützen, as if he was destined to bring this era to an end (April 10, 1888; *KSB* 8.288).

3. I have not attempted to provide a critical account of more than a small number of the vast range of philosophical and scholarly works that comment on Nietzsche's political thought. I hope that the argument of this book and its focus on a specific set of interrelated concepts will help the reader understand that this would have been distracting, as well as almost impossible, given the diversity of those works in focus and orientation. For a sense of this diversity, readers could begin by consulting Tracy Strong, "Nietzsche's Political Misappropriation," in Magnus and Higgins, eds., *The Cambridge Companion to Nietzsche,* 119–47. Strong has also edited the important collection *Friedrich Nietzsche,* a rich assemblage of essays on topics such as responsibility, democratic and aristocratic possibilities in Nietzsche's thought, his relation to critical theory, as well as recent assessments of the general nature of Nietzsche's political project. Two very clear books of a general type but rewarding at several levels are Keith Ansell-Pearson, *An Introduction to Nietzsche as Political Thinker*, and Daniel Conway, *Nietzsche and the Political.* The essays in Jacob Golomb and Robert Wistrich, eds., *Nietzsche, Godfather of Fascism?* attempt to answer or clarify the question posed by the title. In *Leo Strauss and Nietzsche*, Lawrence Lampert offers a sympathetic account of Nietzsche's role in Strauss's thought. *Nietzsche, Power and Politics: Rethinking Nietzsche's Legacy for Political Thought*, Herman Siemens and Vasti Roodt, eds., contains over thirty essays representing recent positions taken by an international group of scholars. Lawrence Hatab provides *A Nietzschean Defense of Democracy*, while Tamsin Shaw explores *Nietzsche's Political Skepticism.* Bernice Glatzer Rosenthal provides an informative set of essays examining Nietzsche's changing meanings in and after the Soviet Revolution in her *Nietzsche and Soviet Culture: Ally and Adversary.* Readers can find a truly comprehensive set of essays and excerpts in *Nietzsche: Critical Assessments,* Daniel Conway, ed. While this

bibliographical note is limited mostly to recent Anglophone accounts, later discussions will involve a number of other relevant texts not cited here.

4. Günter Figal, lecture at Collegium Phaenomenologicum, July 2001.

5. This mistranslation appears in two recent and otherwise generally good scholarly translations of *Beyond Good and Evil* by Judith Norman and Adrian Del Caro, and extends to other passages where the term *Menge* occurs.

6. Cf. Slavoj Žižek, *Living in the End Times*, 98–99. Žižek follows Nietzsche (without acknowledgment) in seeing European Buddhism as a dangerous form of nihilism contrasting with the productive violence of Christian love.

7. The statement is the conclusion of this passage: "Denn wenn die Wahrheit mit der Lüge von Jahrtausenden in Kampf tritt, warden wir Erschütterungen haben, einen Krampf von Erdbeben, eine Versetzung von Berg und Thal, wie dergleichen nie geträumt worden ist. Der Begriff Politik ist dann gänzlich in einen Geisterkrieg aufgegangen, alle Machtgebilde der alten Gesellschaft sind in die Luft gesprengt—sie ruhen allesamt auf der Lüge: es wird Kriege geben, wie es noch keine auf Erden gegeben hat. Erst von mir on giebt es auf Erden *grosse Politik*" (KSA 6.366).

8. Bruno Bosteels, "Nietzsche, Badiou, and Grand Politics: An Antiphilosophical Reading," 213–39.

9. Edmund Husserl, "Foundational Investigations of the Phenomenological Origin of the Spatiality of Nature: The Originary Ark, the Earth, Does Not Move."

10. There seems to be no universal agreement on when to date the beginning of the anthropocene. While some would see its beginning as late as the industrial revolution of the eighteenth century, I suggest seeing it as a process that began with the developments in urban life and agriculture, which our earliest evidence dates to about 10,000–8,000 BCE.

11. Sophocles, *Antigone* ll. 332–75.

12. The phrase, as Nietzsche cites it, is a traditional but somewhat altered version of Tertullian's pronouncement. Cf. Pierre Bühler, "Tertullian: the Teacher of the Credo Quia Absurdum."

13. See Thomas Brobjer, "Notes and Discussions: Nietzsche's Knowledge of Kierkegaard." Nietzsche probably knew nothing of Kierkegaard when writing *Dawn*; however, as Brobjer has shown, he very likely came to at least a general knowledge of his thought a few years later, through Köselitz reading him passages from Georg Brandes's survey of nineteenth-century thought and from his own acquaintance with the writings of Hans Martensen.

14. Cf. Keith Ansell-Pearson, *Viroid Life*, and John Richardson, *Nietzsche's New Darwinism.* Nietzsche's fantasy of his own supposed Polish ancestry could be seen as an attempt at reconstructing such a deep lineage. Polish aristocrats referred to themselves (perhaps more wishfully than accurately) as *sarmats* or Sarmatians, believing themselves to be descendants of a group of nomadic warrior nobles who once inhabited what is now northwestern Iran. So Nietzsche could have imagined a lineage reaching back to Zarathustra's time and place.

15. Nietzsche does not explain how Tertullian, traditionally a Roman citizen from Carthage, anticipates German logic. Perhaps he thinks of the absurdist credo as a one-off that finds its true home among the Germans.

16. The thought is developed intensively and extensively in Peter Sloterdijk, *You Must Change Your Life.*

17. Deleuze and Guattari, *What Is Philosophy?* 102.

18. Peter Bergmann, *Nietzsche: The Last Antipolitical German*, 161–63.

19. Letter to Overbeck, April 30, 1884 (*KSB* 6.497).

20. Vanessa Lemm, "Nietzsche's Great Politics of the Event," 192–93.

21. Alain Badiou, *Casser en deux l'histoire du monde?* Citations are from a still-unpublished translation provided by Bruno Bosteels.

22. Bosteels, 8.

23. Badiou, "Who Is Nietzsche?"

24. The dismissive phrase "so-called world-history" in *BT* indicates that, despite his later regrets that the book is excessively Hegelian, Nietzsche already has serious reservations about this signature Hegelian concept. Robert Gooding-Williams carefully analyzes the conflicting Hegelian and Schopenhauerian tendencies of *BT* in *Zarathustra's Dionysian Modernism*, 112–15. As Gooding-Williams points out, in a quasi-Hegelian vein Nietzsche assigns a special role to the German nation in promising a modern revival of Dionysus; yet, sounding more like Schopenhauer, he contrasts the Dionysian with the history of nations (*BT* 7) and regards historical phases as illusions (*BT* 18).

25. E.g., Robert Gooding-Williams, *Zarathustra's Dionysian Modernism*, 65.

26. E.g., Adrian Del Caro, *Grounding Nietzsche's Rhetoric of Earth*.

27. Gaston Bachelard, *Air and Dreams*; Luce Irigaray, *Marine Lover*.

28. E.g, David Farrell Krell, *Postponements: Woman, Sensuality, and Death in Nietzsche;* Graham Parkes, introduction and notes to his translation of *Thus Spoke Zarathustra*.

29. In the 1880s Nietzsche was a careful reader of Friedrich Ratzel's *Anthropo-Geographie*, which argues—against Kant, Hegel, and others—that mobility rather than permanent attachment to territory is the most general characteristic of humans' relation to territory. See also Günzel, *Geophilosophie*, and Shapiro, "Nietzsche on Geophilosophy and Geoaesthetics."

30. For example, Leo Rauch's translation of Hegel's *Introduction to the Philosophy of History* (1988).

31. Paul Loeb, *The Death of Nietzsche's Zarathustra*, 38–41. Socrates's description in *Phaedo* (108c–113d) of the swampy hollows of the earth as we know it in life should be compared with Nietzsche's description of the morass of human history as sludge (*Schlamm*) and mud or swamp (*Sumpf*) in *D* 18 and his calling nationalistic fetishism of the state's territory *Schollenkleberei*, or being stuck in the mud, *BGE* 241.

32. Ernst Kantorowicz, *The King's Two Bodies,* 89.

CHAPTER TWO

1. E.g., J. C. G. Rohl, From Bismarck to Hitler.

2. See Shapiro, *Earthwards: Robert Smithson and Art after Babel*, 21–58.

3. Hegel discussed this theme in *The Phenomenology of Spirit* in the chapter "The Spiritual Animal Kingdom and Deceit, or 'The Matter in Hand Itself'"; cf. Gary Shapiro, "Notes on the Animal Kingdom of the Spirit."

4. The saying is attributed to Muhammad in a number of the collections of his sayings that have canonical or semicanonical status in Islam. See http://www.usc.edu/dept/MSA

/fundamentals/hadithsunnah/bukhari/052.sbt.html#004.052.210. See also Shapiro, "Assassins and Crusaders: Nietzsche After 9/11."

5. See Horton Harris, *David Friedrich Strauss and His Theology*, 67. Nietzsche also refers to these fierce critics in *UO* I.3, where they are described as hoping to unmask Strauss's "diabolical" principles. This seems to be the first appearance in Nietzsche's writings of topics and themes connected with the idea of the Antichrist, which will eventually become the title of a book first described as the initial contribution toward his *Transvaluation of Values* and then as the whole of that project.

6. Franz Overbeck, *How Christian Is Our Present-Day Theology?*

7. *CWFN* 11.216 (29 [53]). Scholars have tended to neglect Nietzsche's reading of Hegel, and specifically the *Lectures on the Philosophy of World-History*. As early as 1865 he records his reading of Hegel and Strauss together (*KSB* 2.85), and the notebooks from the time of composition of the *Unmoderns* contain extensive quotations from and reflections on Hegel's conception of world-history; see, e.g., *CWFN* 11.212–27 (29 [51–74]).

8. *The Old Faith and the New*, II, 73–79.

9. Ibid., 81.

10. Hegel, *Encyclopedia*, par. 549.

11. *CWFN* 11.227 (29[74]).

12. Hegel, *Introduction to the Philosophy of History*, 89–91.

13. Fukuyama, 262.

14. Ibid., 329.

15. I am summarizing elements of Hegel's account of "The German World." See *Philosophy of History*, 341–411.

16. Cf. Anthony K. Jensen, "The Rogue of All Rogues: Nietzsche's Presentation of Eduard von Hartmann's *Philosophie des Unbewussten* and Hartmann's Response to Nietzsche."

17. Eduard von Hartmann, *Philosophy of the Unconscious*, cf. esp. vol. 2, 1–45 and vol. 3, 94–119. See also Gary Shapiro, "Nietzsche's Modernity: On Robert Gooding-Williams's *Zarathustra's Dionysian Modernism*."

18. Clark Blaise, *Time Lord: Sir Sandford Fleming and the Creation of Standard Time*; chap. 5 is titled "The Decade of Time, 1875–85" 69–90.

19. Philosophical parody seems to have attained a certain height in the post-Hegelian nineteenth century, and the target is typically Hegel (or the Hegelians). Perhaps its greatest masterpiece is Kierkegaard's pseudonymous *Concluding Unscientific Postscript*, which parodies a Hegel treatise. Also pseudonymous is Bruno Bauer's *The Trumpet of the Last Judgment Against Hegel, the Atheist and Antichrist*, which disingenuously purports to unmask Hegel's atheism while implicitly suggesting that this is his most viable philosophical position. Bauer's authorship was discovered, and he was consequently unable to hold a regular academic position. It is unlikely that Nietzsche knew either work when he was composing *UO* II, although he later became familiar with some of the writings of Bauer and indirectly with Kierkegaard (on Bauer and Nietzsche see Karl Löwith, *From Hegel to Nietzsche* 296–97, 3339–46; for Nietzsche and Kierkegaard, see Thomas Brobjer, "Nietzsche's Knowledge of Kierkegaard." Nietzsche's notes for his essay on Strauss contain a passage suggesting that *The Old Faith and the New* was a parody meant to disgrace Strauss (*CWFN* 11.161).

20. Eduard von Hartmann, *Philosophy of the Unconscious*, (London: Kegan Paul, 1931), vol. II: 1–45, vol. III: 94–119.

21. Gary Shapiro, *Earthwards: Robert Smithson and Art after Babel*, 21–58.

22. Stanley Cavell, *Conditions Handsome and Unhandsome: The Constitution of Emersonian Perfectionism*, 48–54. Nietzsche's fascination with Emerson is already evident in his 1862 essays on "Fate and History" and "Freedom of Will and Fate."

23. When we read "analysis" in terms of its Greek etymological associations with untying, and "deconstruction" as a translation of Heidegger's *De-struktion* or de-structuring, the concepts seem much more closely related than in textbook cliché and journalistic opinion.

24. For Badiou's thoughts on the suturing of philosophy (to science, politics, art, or love), see *Manifesto for Philosophy*, 61–67.

25. Cf. Bergmann, Nietzsche:"The Last Unpolitical German," 87–88, 95–97.

26. Deleuze and Parnet, *Dialogues*, 13.

27. Deleuze, "Nomad Thought," in *The New Nietzsche*, 149.

28. Kant, Preface to first edition of *Critique of Pure Reason*, 7 (Akademie edition, ix).

29. Cited in an 1882 notebook, *KSA* 9.667; from Emerson, "History," 17.

30. Emerson, "History," 7. Nietzsche's reception of Emerson has received a good deal of attention. See George Stack, *Nietzsche and Emerson: An Elective Affinity*, for a general account. Stanley Cavell has explored Nietzsche's absorption of Emersonian perfectionism in a number of books and essays such as *This New Yet Unapproachable America: Lectures after Emerson after Wittgenstein*. Stephan Günzel notes Nietzsche's reading of a continuation of the Emerson passage on "spiritual nomadism" in *Geophilosophie: Nietzsches philosophische Geographie*, 55–56.

31. Emerson, "History," 17.

32. Marx and Engels, in the first few pages of *The German Ideology*, speak with deep sarcasm about "the industrialists of philosophy" who have attempted to market their various attempts to exploit absolute Spirit; Karl Marx, *Selected Writings*, 104–5. Nietzsche could not have known this work, which was unpublished until 1932.

33. Nietzsche makes several admiring references to the Jesuits, showing respect for their organization and discipline, and he is fairly consistent in preferring Catholicism to Protestantism.

34. Shapiro, *Nietzschean Narratives*, 17–20.

35. Schopenhauer, "On Philosophy at the Universities," in *Parerga and Paralipomena*, vol. I, 153.

36. "On Philosophy at the Universities," 152.

37. Ibid.," 173. This *Jetztzeit*, of course, must be distinguished from Walter Benjamin's use of the term to designate messianic time.

38. Hegel, *Philosophy of Mind* (Encyclopedia III), par. 549.

39. Hegel, *Introduction to the Philosophy of History*, 42.

40. Hegel, *Philosophy of Mind (Encyclopedia* III*)*, par. 549. Here Hegel's *Encyclopedia* translator (A.V. Miller) tends to use the word "nation," possibly to avoid the racist associations that the German *Volk* acquired under the Nazis; the term has its roots in the romanticism of Hegel's time, going back to writers like Herder. However, we need to recall that much of Hegel's supposed "rationalism" was developed as a critique of such "romanticism." Our word "nation" can work as a translation of *Volk* if we recall its etymological ties to nature and natality, coming to

be or birth, with implications of common origin, kinship, and attachment to territory or earth. Yet we sometimes use "nation" to designate a political state. The nineteenth century invented the idea of the "nation-state," as the combination of a people or folk and their political form. When Hegel contrasts the mere *Volk*, which is not a subject of history, with the states, which are, as in this *Encyclopedia* claim: "A *Volk* with no state formation (a mere nation/*Nation*) has, strictly speaking, no history—like the *Völker* which existed before the rise of states and others which still exist . . . as mere lineages (*Nationen*)" (par. 549, p. 279), Miller translates the phrase "wild nations" as "still exist in a condition of savagery" and otherwise confuses important distinctions in Hegel among state, people, and nation.

41. Badiou, *Manifesto for Philosophy*.

42. Hegel, *Philosophy of History*, 341.

43. For an effective assemblage of such comments, see J. P. Stern's "Introduction" to Nietzsche, *Untimely Meditations*, trans. R. J. Hollingdale, xxvi–xxviii.

CHAPTER THREE

1. Carl Schmitt, *Political Theology*, 5.

2. The literature is vast; a good place to start is Jacob Golomb, ed., *Nietzsche and Jewish Culture*.

3. Cited in Erich Heller, "Introduction" to *Human, All-Too-Human,* xi.

4. As usual, we should be cautious in translating *Mensch*, which is generally best rendered as "human," not as "man." I deviate from this practice a number of times in the case of "the last man," a phrase that has established itself in English.

5. The "artificial Catholicism" to which Nietzsche refers would seem to be the situation that resulted from the principle of *cuius regio eius religio*, first announced in the Peace of Augsburg (1555) and established more widely and firmly by the Peace of Westphalia (1648); this allowed the ruler of each state, including the multiplicity of German states that preceded Bismarck's *Reich*, to determine the state's religion.

6. Hegel, *Philosophy of History*, 358.

7. Schmitt, *Political Theology*, 5.

8. See, e.g., Simon Critchley, *Infinitely Demanding: Ethics of Commitment, Politics of Resistance*, "Crypto-Schmittianism—The Logic of the Political in Bush's America," 133–48.

9. See Giorgio Agamben, *State of Exception*.

10. The publisher of Nietzsche's *HAH*, Ernst Schmeitzner, sent a copy to Bismarck, who answered Schmeitzner with a thank you note that has not been preserved. In a note to Schmeitzner, Nietzsche said that if Bismarck had actually read it, he would have thrown it at the wall (*KSB* 2.345).

11. Agamben, *State of Exception*, 16–18, citing Carlo Fresa.

12. In a study of *Nietzsche's Political Skepticism*, Tamsin Shaw offers a reconstruction of Nietzsche's argument against the legitimacy of the state that relies on cognitivist or epistemic principles. In her reconstruction, Nietzsche argues that any normative claim by the state would require its knowledge of relevant truths. Not only are any such claims to truth questionable, but the demand for truth is also in irremediable tension with its need to command general

agreement (139). Shaw comments (23) in some detail on Nietzsche's account in *HAH* 472, discussed above, of the gradual "withering away" of the state; however, she does not explicitly address Nietzsche's recognition of the importance of the contemporary importance of the state of exception, although her study involves a rich discussion of the Bismarckian context of Nietzsche's political thought. Although she discusses Carl Schmitt with reference to other questions, she does explicitly juxtapose their thought on the state. Both see political sovereignty in any strong sense as "types" or structural variations of divine sovereignty, and draw the appropriate conclusions, given their relation to God.

13. See James Porter, *Nietzsche and the Philology of the Future*, 43. See also *BGE* 48 on Comte's crypto-Catholic tendencies.

14. *CWFN* 11, 304.

15. For a striking account of the politics of philosophy in the United States since the beginning of the Cold War, see John McCumber, *Time in the Ditch.*

16. *KSA* 8.473–74; cf. *Dawn* 454.

17. Robert Gooding-Williams, *Zarathustra's Dionysian Modernism*, 128–31.

18. *KSB* 8: 222.

19. Luce Irigaray, *Marine Lover.*

20. Ernst Bertram, *Nietzsche: Attempt at a Mythology.*

21. A very comprehensive account of Nietzsche's concern with the earth, with constant reference to Deleuze and Guattari, is to be found in Stephan Günzel, *Geophilosophie: Nietzsches philosophische Geographie*; for detailed analyses of the poetic and psychic sense of the landscapes of Zarathustra, see 241–63. Adrian Del Caro, *Grounding the Nietzsche Rhetoric of Earth*, undertakes a systematic linguistic and rhetorical study. Graham Parkes, "Staying Loyal to the Earth: Nietzsche as an Ecological Thinker," is a brief and illuminating treatment.

22. Lawrence Lampert does not mention the theme at all in his Zarathustra commentary, *Nietzsche's Teaching*, while observing that Zarathustra's rhetoric here is extreme and inflated. Robert Gooding-Williams does not thematize the question of death in the section of *Zarathustra's Dionysian Modernism* devoted to this chapter.

23. Nietzsche's thematics of blood and scarlet in Zarathustra's vision of the murderer and his judges should be compared with some of Derrida's analyses in *The Death Penalty: Vol. 1*, esp. the ninth session, 218–42.

24. Michel Foucault, The History of Sexuality: Vol. 1, 133–59.

25. An important study of the danger posed by state-driven rationalistic, top-down planning and organization is James Scott's *Seeing Like a State: How Certain Schemes to Improve the Human Condition Have Failed.* Scott's analysis resonates both with Nietzsche's critique of abstract measurement of population and its conditions and Deleuze and Guattari's distinction between spaces that are striated (by centrally imposed geometrical lines) and smooth (in which paths and mobility are negotiated "on the ground" in terms of local variations and circumstances); *A Thousand Plateaus*, chap. 14, "The Smooth and the Striated," 474–500. Like Emerson, Nietzsche, and Deleuze/Guattari, Scott takes as a fundamental point of departure the state's enmity toward "people who move around," 1.

26. Peter Sloterdijk, *You Must Change Your Life*, 343–45.

27. Foucault, "La force de fuir," in *Dits et écrits*, vol. 2, 401–05.

28. The very useful notes to Graham Parkes's translation of *Zarathustra* point out some of the possible sources of the peculiar framing story that Nietzsche employs in this chapter; see *Thus Spoke Zarathustra*, 303–04. Also worth noting is Ratzel on volcanic mythology, *Anthropo-Geographie*, 396–98. Among other "superstitions," Ratzel mentions the belief that volcanic lava contains gold, which may have inspired Zarathustra's saying that the heart of the earth is of gold.

29. For the Rousseau allusion, see *UO* III.4 and Lampert, *Nietzsche's Teaching*, 332, n.79.

30. Cf. Gary Shapiro, "Dogs, Domestication, and the Ego." I suggest that a dog in Nietzsche often represents the ego as limited by excessive individuation—as in Nietzsche's jotting, "And wherever I climb, my dog follows me everywhere; he is called 'ego'" (*KSA* 10.165; 4[188]).

31. Any careful reading of *Zarathustra* will have to take into account Nietzsche's reworking of the Empedocles legend, which is evident in his notes for the text. See David Farrell Krell, *Postponements: Woman, Sensuality, and Death in Nietzsche.* Carl Jung discusses an intriguing literary parallel for this chapter, from a story retold by Justinus Kerner, that Nietzsche may have been drawing on unconsciously. Jung, *Nietzsche's Zarathustra*, vol. 2, 1217–18.

32. Deleuze and Guattari, *A Thousand Plateaus*, 40.

33. Like Zarathustra, Nietzsche welcomes the dice game of the gods (alluding to the sayings of Heraclitus), including earthquakes, which remind us that all foundations are fragile and contingent. In February 1887, he writes to Reinhard von Seydlitz, after surviving a severe earthquake in Nice: "We are now living in the interesting expectation of *perishing*—thanks to a well-meaning earthquake that has everyone here baying at the moon, and not just the hounds. What pleasure it is when these ancient houses rattle over our heads like coffee grinders!" (*KSB* 8, 31–32). "Well-meaning" earthquakes are salutary messages confirming the earth's process and the ad hoc character of the arrangements of the *Menschen-Erde* (Nietzsche would have welcomed plate tectonics, which demonstrates that the earth is surprisingly more mobile than even the geologically informed nineteenth century was able to guess).

34. Although I once read the fourth part as a sequel to the first three, I now agree with Paul Loeb (*The Death of Nietzsche's Zarathustra*) that Part IV is best understood as an interlude in the narrative of Parts I–III. In either of these readings, Part IV can be read in terms of its similarities to the ancient Greek satyr play that accompanied three tragedies.

35. *What Is Philosophy?*, 102.

36. Ibid., 85.

37. Ibid., 93.

38. Ibid., 91.

39. Gayatri Chakravorty Spivak, "Can the Subaltern Speak?"

40. Michel Foucault, "Nietzsche, Genealogy, History," in *Language, Counter-Memory, Practice.* Robert Holub argues that Foucault has exaggerated the consistency of Nietzsche's alleged *Ursprung/Herkunft* distinction in his essay "Reading Nietzsche as Postmodernist." However, Holub shows the limits of his attempt to reconstruct a more "dogmatic" Nietzsche when he writes without qualification, "Who can read the *Genealogy* and still believe that Nietzsche does not consider the slave morality of good and evil inferior to the good-and-bad system of the blond beasts?" (Holub, 261). Yet Nietzsche claims in a well-known analysis that human inwardness was produced by the tensions of civilization, in which former blond beasts adopt elements of "slave morality" in self-criticism, since their aggressive instincts could no longer find full satisfaction

in action. He says that the existence of an animal psyche turned against itself brought about on earth something so new, profound, enigmatic, contradictory, and *full of future* that the aspect of the earth changed essentially" (*GM* II.16). As Nietzsche implies, there was no future for humans or the earth in remaining within the simple duality of master and slave morality that preceded what he calls "the great dice throw" of this event. Holub criticizes an article in which I provisionally adopted Foucault's reliance on the *Ursprung/Herkunft* distinction in order to point out residues of "origin" thinking in his presentation of the *Genealogy*; "Translating, Repeating, Naming: Foucault, Derrida, and *The Genealogy of Morals.*"

41. Gilles Deleuze and Claire Parnet, *Dialogues*, 131–32.

42. Hegel, *Philosophy of Right*, 13. Hegel's famous saying is, "When philosophy paints its gray in gray, then a shape of life has grown old. By philosophy's gray in gray it cannot be rejuvenated but only understood. The owl of Minerva spreads its wings only with the falling of the dusk."

43. *Hegel's Aesthetics*, vol. II, 888–901.

44. Deleuze and Guattari, *A Thousand Plateaus,* chap. 11.

45. Ratzel, *Anthropo-Geographie*, 438.

46. In several passages, Nietzsche seems to implicitly distinguish *Menge* and *Masse* in terms of the greater diversity of *Menge*. Note esp. *GS* 149, where Nietzsche says that in Greece "there must have been a multitude (*Menge*) of diverse individuals," contrasting this, later in the aphorism, with the homogeneity of the *Masse*. While his usage is not completely consistent, the prevailing tendency in Nietzsche's texts is almost inevitably to associate the *Masse* with the relatively uniform (and often the *Heerde*). A crucial passage is 7.642, where Nietzsche writes, "Statistics prove that there are laws in history. Indeed, it proves how common and disgustingly uniform the mass (*Masse*) is. You should have tried statistical analysis in Athens for once! The lower and more non-individual the mass (*Masse*) is, the statistical laws are that much stronger. If the multitude (*Menge*) is finer and nobler, the law goes to the devil." See also 4.18, 7.119, 9.462, 12.96. For a recent ontological and political analysis of the concept of multitude, see Michael Hardt and Antonio Negri, *Multitude: War and Democracy in the Age of Empire*, esp. 99–102.

Inspired by Spinoza, Hardt and Negri argue that multitude must be distinguished from concepts such as crowd, masses, mob, and rabble: "The crowd or the mob or the rabble can have social effects—often horribly destructive effects—but cannot act of their own accord. That is why they are so susceptible to external manipulation. The multitude designates an active social subject, which acts on the basis of what the singularities have in common" (100). Alain Badiou, drawing on set theory (*Mengenlehre*), understands the multiple as a fundamental ontological and political category; see, e.g., *Being and Event*, esp. 104–11.

47. Deleuze and Parnet, *Dialogues,* 142.

CHAPTER FOUR

1. Carl Schmitt, "Nomos—Nahme—Name," in *The Nomos of the Earth in the International Law of the Jus Publicum Europaeum*, trans. G. L. Ulmen (Telos Press, 2006) 336–50; see esp. 347–50.

2. Schmitt, "*Nomos—Nahme—Name*," 349.

3. *Das Problem der Wartenden*:

> Es sind Glücksfälle dazu nöthig und vielerlei Unberechenbares, dass ein höherer Mensch, in dem die Lösung eines Problems schläft, noch zur rechten Zeit zum Handeln kommt—"zum Ausbruch," wie man sagen könnte. Es geschieht durchschnittlich *nicht*, und in allen Winkeln der Erde sitzen Wartende, die es kaum wissen, in wiefern sie warten, noch weniger aber, dass sie umsonst warten. Mitunter auch kommt der Weckruf zu spät, jener Zufall, der die "Erlaubniss" zum Handeln giebt,—dann, wenn bereits die beste Jugend und Kraft zum Handeln durch Stillsitzen verbraucht ist; und wie Mancher fand, eben als er 'aufsprang,' mit Schrecken seine Glieder eingeschlafen und seinen Geist schon zu schwer! "Es ist zu spat"—sagte er sich, ungläubig über sich geworden und nunmehr für immer unnütz.—Sollte, im Reiche des Genie's, der "Raffael ohne Hände," das Wort im weitesten Sinn verstanden, vielleicht nicht die Ausnahme, sondern die Regel sein?—Das Genie ist vielleicht gar nicht so selten: aber die fünfhundert *Hände*, die es nöthig hat, um den καιρός, "die rechte Zeit"—zu tyrannisiren, um den Zufall am Schopf zu fassen! (*BGE*, 274).

4. Cf. Lawrence Lampert, *Nietzsche's Task: An Interpretation of Beyond Good and Evil*, and Douglas Burnham, *Reading Nietzsche: An Analysis of Beyond Good and Evil*. Acampora and Pearson suggest that the narrative continuity of Nietzsche's book breaks off at this point in the final chapter, constituting an interlude or *Zwischenspiel* of more personal and episodic observations before the final two aphorisms; Christa Acampora and Keith Ansell Pearson, *Nietzsche's Beyond Good and Evil*, 204–05.

5. James Scott, *Seeing Like a State*, 343. Scott remarks "how routinely planners ignore the radical contingency of the future." As he shows in rich detail, the modern state has all too frequently failed at ambitious projects while undermining itself through a naïve temporal ontology. Scott's own alternative political virtue, drawing on Marcel Detienne and Jean-Paul Vernant, is *mētis*, the skill, disposition, or art of "cunning intelligence," which adapts to the very specific circumstances and textures of a complex situation (as in the arts of navigation and medicine, as Aristotle describes them in an account of *phronēsis*). While Nietzsche might praise Scott's recognition of radical contingency and the need for a creative response to the unpredictable, he might ask whether Scott's "empiricism" limits his ability to see a *kairos* or Machiavellian *occasio* when it rushes by. Similarly, Scott might ask whether Nietzsche's expectation of a "great event" has disturbing similarities with the "will to power" of the authoritarian planning state; this, he might very well suggest, points to Nietzsche's having a certain deficit in *mētis*. I will be arguing that the outlines of a Nietzschean response to such a critique can be found in "What Is Noble?" (*BGE*).

6. Deleuze, *Difference and Repetition*, 157–67, 195–204.

7. Acampora and Pearson, 209–11.

8. For a discussion and reproduction of a paradigmatic ancient image of *kairos*, see Erwin Panofsky, *Studies in Iconology: Humanistic Themes in the Art of the Renaissance*, 71, 93, plate XXI.

9. Nietzsche certainly knew Machiavelli's discussion of luck (fortune) and occasion (opportunity) in *The Prince*, in which he develops the theme of balance and imbalance. Some rulers prudently attempt to keep things close to a constant balance. They may succeed in this, but

they will never have the greater achievements of the more daring ones who can see and act on the promising imbalances of the occasion. There comes a time when only bold action is successful, so "the cautious man when it is time to act boldly, does not know how, and comes to grief." While the success of either the cautious or the bold man depends on the times, Machiavelli concludes the discussion with this notorious observation: "It is better to be rash than timid, for Fortune is a woman, and the man who wants to hold her down must beat and bully her." Note that from earlier classical times the gender of the personified figure of Occasione, the Latin equivalent of kairos, had changed from masculine to feminine. In a poem that appears in Machiavelli's works, but is actually his loose translation of a fourth-century Latin epigram of Ausonius, Occasione has winged feet and an extremely long head of hair that hangs frontally, with the traditional bald spot behind (*The Prince*, chap. xxv, 72–73 and 138).

10. See Morton Smith, *Jesus the Magician*, and Geza Vermes, *Christian Beginnings*.

11. Cavell, *Conditions Handsome and Unhandsome*. Cf. James Conant, "Nietzsche's Perfectionism: A Reading of *Schopenhauer as Educator*," in *Nietzsche's Postmoralism*, 181–257.

12. G. E Lessing, *Emilia Galotti*, 7–8.

13. Yirmiyahu Yovel, "Nietzsche and the Jews: The Structure of an Ambivalence," 132.

14. Nevertheless, two recent commentaries on *BGE*, by Lawrence Lampert and Douglas Burnham, which are valuable attempts to understand the book as a whole, give scant attention to Nietzsche's theme of the multitude and its political significance. While both offer discussions of the herd and herd morality, neither notes the place of "herd" within a spectrum of distinguishable concepts designating various types of groups.

15. See *Menge* in Grimm's *Wörterbuch*.

16. Some further linguistic evidence concerning Nietzsche's emphasis on the diversity and multiplicity of the *Menge* may be found in mathematical discourse. The *Menge* is the sheer multiple, a set of diverse elements. While the set may have a definition (the set of all odd numbers, all engaged in productive work, all who live in Europe, etc.), they are internally diverse. The mathematician Georg Cantor was developing set theory (*Mengenlehre*) around the time Nietzsche was describing a world of limitless multiplicity. I find no evidence that Nietzsche knew of Cantor's discoveries. However, he was clearly interested in analogous concepts, such as the Darwinian idea of a population, as opposed to an essentially defined species (e.g., *GS* 1, 149).

17. Another crucial passage: "Statistics prove that there are laws in history. Indeed, it proves how common and disgustingly uniform the mass [*Masse*] is. You should have tried statistical analysis in Athens for once! The lower and more non-individual the mass [*Masse*] is, the statistical laws are that much stronger. If the multitude [*Menge*] is finer and nobler, the law goes to the devil" (*KSA* 7.642; cf. *KSA* 4.18, 7.119, 9.462, 12.96).

18. J. W. Goethe, *Faust*, 73–75.

19. One apparent counterexample: "It is a great achievement when the *grossen Menge* (people of all kinds who lack depth or have speedy bowels) have finally had the feeling bred into them that they cannot touch everything . . ." (263). Here the phrase is *grossen Menge*, however, which suggests a more inclusive group than the *Menge* as such; most translators correctly choose "masses."

20. Elizabeth Struthers Malbon, *In the Company of Jesus: Characters in Mark's Gospel*. This is a careful reading of Mark with reference to the roles of disciples and crowd; see esp. chap. 3, "Disciples/Crowds/ Whoever: Markan Characters and Readers," 70–99. The Greek texts of the

synoptic gospels typically speak of the *ochlos*; John and Acts more frequently employ *plethos*, but Luther usually translates both as *Menge*. See Matthew 9:33, 14:5, 21:46; Mark 5:31, 8:1–2; Luke 2:13, 8:45, 12:1, 22:6, 22:47, 23:1; John, 5:13, 21:6; Acts 4:32, 6:5, 15:12, 23:7. See also the Jester's warning to Zarathustra that unnamed powerful figures consider him a danger to the multitude and the Jester's suggestion that it is only their laughter that has allowed Zarathustra to escape the town alive (*Z* P 8).

21. In this aphorism, Nietzsche sometimes speaks of "*grosse Menschen*" and sometimes of specifically gendered "*grosse Männer*."

22. Cf. Étienne Balibar, "Spinoza, The Anti-Orwell: Fear of the Masses." Balibar does not distinguish among terms such as "masses" and "multitude" and seems to use them interchangeably.

23. See Christa Acampora, "Demos Agonistes Redux: Reflections on the *Streit* of Political Agonism," and works cited there.

24. Cf. Hubert Cancik, "'Mongols, Semites, and the Pure-bred Greeks': Nietzsche's Handling of the Racial Doctrines of His Time."

25. Giorgio Agamben, *The Time That Remains*, 68–69.

26. Hegel, *Philosophy of World History*, 347–54.

27. I leave "*Mensch*" untranslated to preserve the etymological point without the gendered English term "man"; although *Mensch* is grammatically masculine, it is used to designate humans of both genders and is distinguished from the more specific term *Mann*.

28. Shapiro, *Alcyone*.

29. Cf. Heidegger, "Anaximander's Saying," in *Off the Beaten Track* (*Holzwege*), 242–81, and Shapiro, "Debts Due and Overdue: Beginnings of Philosophy in Nietzsche, Heidegger, and Anaximander," 358–75.

30. Charles Kahn, "Anaximander's Fragment: The Universe Governed by Law," 106–9.

31. Michel Foucault, "Nietzsche, Genealogy, History."

32. For a recent ambitious account of human history in terms of debt and credit, see David Graeber, *Debt: The First Five Thousand Years*.

33. Deleuze and Guattari, *Anti-Oedipus: Capitalism and Schizophrenia*, 139–271

34. The status of these who are both inside and outside the community might be compared to Agamben's understanding of the "saving remnant" of which Paul speaks, to Badiou's concept of those excluded from a society's conventional "we," the *sans-papiers*; cf. *Infinite Thought*, 68–78. Cf. also Jacques Rancière's idea of "the part of those who have no part," as in "Who Is the Subject of the Rights of Man?"

CHAPTER FIVE

1. A notable exception is Loeb, *The Death of Nietzsche's Zarathustra*.

2. On *Zarathustra* as a landscape poem, see chapter 3. Despite the profusion of close readings of *Zarathustra* in recent decades, the reader will generally search in vain for "garden" in the index of any Anglophone book devoted to *Zarathustra*. Similarly, scant attention has been given to the thought that the world awaits Zarathustra as a garden. Lawrence Lampert writes of Zarathustra's openness to the natural, as exemplified by his ability to understand the speech of his animals, while simply mentioning the one point on which he explicitly agrees with

EPICURUS

the animals: the world awaits him as a garden; cf. *Nietzsche's Teaching*, 211–23. Like Lampert, Gooding-Williams offers a careful, nuanced reading of the interchange between Zarathustra and the animals, but ignores the invitation extended by the latter to welcome the world as a garden; *Zarathustra's Dionysian Modernism* 250–53. Such examples could be multiplied, and Internet searches will yield only a few writers who have had something to say about Nietzsche and the topos of the garden; these almost all have to do with the analogy or metaphor of self-cultivation of the drives (as in passages from *Dawn*, discussed below) or with Nietzsche's sympathetic reading of Epicurus and his "garden philosophy" (also discussed below).

Several readers also note the description of marriage as a garden-happiness of the earth (*Z* I.20). See also *HAH* II.278, which treats the analogy in negative fashion, suggesting that proscribing the "discontented, atrabilious and sullen" from reproducing "could magically transform the earth into a garden of happiness.—This proposition belongs in a practical philosophy for the female sex." Del Caro observes that Nietzsche here seems less than completely sincere, since he apparently deviates from his more usual tendency to seek long-term solutions for long-term problems; *Grounding the Nietzsche Rhetoric of Earth*, 84. Other than this reference, Del Caro barely mentions the garden theme, which is similarly almost absent in Günzel, *Geophilosophie*. A collection of essays *Nietzsche and "An Architecture of Our Minds,"* ed. Alexandre Kostka and Irving Wohlfarth, contains a number of studies of Nietzsche's thinking about architecture, including, marginally, landscape architecture, but offers no specific reading of this and related passages. The most comprehensive account of some of the relevant texts (including letters) is Tilman Buddensieg, *Nietzsches Italien: Städte, Gärten, Paläste*; the focus, as the title suggests, is on Nietzsche's Italian travels and residencies. I have discussed some related topics in "Territories, Landscapes, Gardens: Toward Geoaesthetics" and "Nietzsche on Geophilosophy and Geoaesthetics," 477–94. Chenxi Tang has written an important study that clarifies the romantic background of Nietzsche's landscape sensibilities. In *The Geographic Imagination of Modernity: Geography, Literature, and Philosophy in German Romanticism*, Tang discusses the Schlegels, Novalis, Hölderlin and others, showing that their landscape aesthetics was informed by geography and geology. While not explicitly discussing Nietzsche, the book is a useful corrective to accounts of the romantics that ignore the scientific and empirical dimensions of their landscape writing. Like Ratzel's *Anthropo-Geographie*, it should be read along with such phenomenological analyses of Nietzsche as Bachelard and Irigaray.

3. Cf. Loeb, *The Death of Nietzsche's Zarathustra*, 167–68.

4. Does *Zarathustra* end with Part III or Part IV? As noted earlier, the most intriguing recent answer to the question, taking Part IV to be an interlude within the action of Part III, is given by Paul Loeb, *The Death of Nietzsche's Zarathustra,* 85–97.

5. On the cosmic dimension of landscape in "The Seven Seals," see Shapiro, *Alcyone: Nietzsche on Gifts, Noise, and Women*, 134–36.

6. Martin Heidegger, "The Origin of the Work of Art," in *Poetry, Language, Thinking*. Heidegger seems not to have said anything concerning the theme of the garden (perhaps with the partial exception of his topos of the peasant farmstead), despite his musings on Greek and Swabian landscapes and his obsession with Hölderlin's geographically oriented poetry.

7. Hegel, *Lectures on Aesthetics*, vol. II, 702.

8. Paul Kristeller, "The Modern System of the Arts."

9. Another more recent book by Larry Shiner gracefully traces the various approaches to and evasions of aesthetics beginning with the Greeks. Cf. *The Invention of Art: A Cultural History.*

10. Cited by John Dixon Hunt and Peter Willis, *The Genius of the Place: The English Landscape Garden 1620–1820*, 11.

11. For a recent set of essays exploring the philosophical meanings of gardens and gardening, see *Gardening: Cultivating Wisdom*, ed. Dan O'Brien.

12. Elizabeth Barlow Rogers, *Landscape Design: A Cultural and Architectural History*, 314.

13. Jacob Burckhardt, *Der Cicerone: Eine Anleitung zum Genuss der Kunstwerke Italiens.*

14. Cf. Deleuze, "A New Cartographer," in *Foucault*, 23–46. "The diagram is no longer an auditory or visual archive but a map, a cartography that is coextensive with the whole social field," 34.

15. I have argued elsewhere that Foucauldian archaeology draws on Nietzschean inspirations and models. The concept of the diagram as it appears in *Discipline and Punish* can be read as a formal variation on Nietzsche's analysis of the complex play of perspectives in the Greek theater (cf. *BT* 8); Shapiro, *Archaeologies of Vision: Foucault and Nietzsche on Seeing and Saying* 135–43, 293–301.

16. Cf. Rogers, *Landscape Design*, 125–49.

17. Oswald Spengler, *The Decline of the West*, vol. I, 241.

18. Ibid., 254.

19. Uvedale Price, *Essays on the Picturesque*, vol. I.

20. John Dixon Hunt, "'Ut Pictura Poesis': The Garden and the Picturesque in England (1710–1750)," in *The History of Garden Design: The Western Tradition from the Renaissance to the Present Day*, eds. Monique Mosser and George Teyssot.

21. Foucault, *Discipline and Punish*, 201–2.

22. Joseph Addison, *The Spectator* 1712, no. 412.

23. Kant, *Critique of Judgment*, section 51.

24. Ibid.

25. On the complex interchanges involving productive land use, representation, and garden construction, see the following studies: Paul Shepard, *Man in the Landscape: a Historic View of the Esthetics of Nature*; Gina Crandell, *Nature Pictorialized: "The View" in Landscape History*; Shapiro, "Territories, Landscapes, Gardens."

26. For Schopenhauer's discussions of natural beauty, parks, gardens, and related subjects, see Arthur Schopenhauer, *The World as Will and Representation*, vol. I, 217–20; vol. II, 374–75, 381, 403–05.

27. The literature of the last fifty years or so on the meaning of place is enormous and multifarious. I cite Casey's work here because, I think, it captures especially well the sense of possibility and futurity that places may have, as well as clarifying several distinctions between space and place. A representative but very partial selection of other major contributions might begin with the following: much of Heidegger's work, including the analysis of *Dasein*'s spatiality in *Being and Time* and such essays as "Building Dwelling Thinking," in *Poetry, Language, Thought*; Maurice Merleau-Ponty, *The Phenomenology of Perception*; Bachelard, *The Poetics of Space*; Henri Lefebvre, *Production of Space*; Peter Sloterdijk's series *Spheres*, including *Bubbles* and *Globes* and the still-untranslated *Schäume* or *Foam.*

28. Edward S. Casey, *Getting Back into Place: Toward a Renewed Understanding of the Place-World*, xiv.

29. Casey, *Getting Back into Place*, 111.

30. Ibid., 168–69. On being and dwelling, see both Heidegger, "Building Dwelling Thinking," and Albert Hofstadter, "Being: The Act of Belonging," in *Agony and Epitaph*, 199–257.

31. For Sloterdijk, who cites Nietzsche frequently, see n. 27. For Gernot Böhme, see *Atmosphäre: Essays zur neuen Ästhetik.*

32. Gaston Bachelard offers a striking reading of "Nietzsche and the Ascensional Psyche" in *Air and Dreams*, 127–60. Bachelard maintains that Nietzsche's poetic imagination (and consequently his philosophical concepts) are not oriented to the earth in an elemental sense as contrasted with water, fire, and air. He understands Nietzsche's cosmos as a cosmos of the heights, emphasizing Nietzsche's strong preference for the cold air and silence of the heights (130). The analysis is illuminating but partial. Yet for Nietzsche, the earthbound garden is a desirable perch from which to experience the surrounding atmosphere—which is not just cold, empty space; it is open to the sky and its imaginative depths. Bachelard endorses Nietzsche's "aestheticization of ethics" (144) and acutely describes his imagination of nature, as in *Zarathustra* and his poetry, as an "experimental physics of the moral life (149).

Perhaps if Bachelard had given more attention to the political dimension of the summons to be true to the earth, he would have seen that it is politics as well as ethics and morals that are aestheticized. That reading would involve seeing Nietzsche's struggle with world-history as crucial for understanding his hopes for the human-earth. Nietzsche's projection of a future human-earth that would be a great tree is a proposal for transforming the earth into an aerial garden, an idea that nicely combines the "ascensional" with the garden as a model for living "on" earth. No doubt, Nietzsche would have partially confirmed Bachelard's analysis by agreeing with the common advice that height is the most neglected dimension of garden design.

33. The power of stage design in that of gardens is noted by Hunt in "'Ut Pictura Poesis,'" 231–33, and by Crandell, *Nature Pictorialized*. Böhme does not seem to have noticed this connection, although he does comment on the important eighteenth-century German treatise on garden design by Hirschfeld; cf. *Atmosphäre*, 36–37.

34. Zarathustra's plantation is presented as windswept, gnarled, and braving the elements, rather than as a geometrically structured "industrial" forest project. In this connection, it is interesting that in *Seeing Like a State*, Scott's first and paradigmatic illustration of how top-down, overly rationalized planning goes wrong, is drawn from the misadventures of "scientific forestry." He observes that "the forest as a habitat disappears and is replaced by the forest as an economic resource to be managed efficiently and properly," 13.

35. On landscape as face, see Gilles Deleuze and Félix Guattari, *A Thousand Plateaus*, 172–73.

36. See Buddensieg, *Nietzsches Italien*, 27–57, for a very suggestive account of Nietzsche's response to the architecture and gardens of Genoa, with reference to letters and other sources. I can testify that the Villetta di Negro, the small park just across the street from Nietzsche's rooms in Salita delle Battistine 8, is an extraordinary work of landscape architecture. With steep declivities, a waterfall, and spectacular views of Genoa and the bay, it is understandable that Stendhal called it one of the most picturesque places in Italy. In letters Nietzsche calls it his

garden, and offers this description: "Very near is a charming garden, open to the public, with a powerful, forest-like greenery (also in *winter*), waterfalls, wild animals and birds and magnificent distant vistas of the sea and mountains—all in a very small space" (*KSB* 6.151).

37. Jacob Burckhardt, *Der Cicerone,* 379–80.

38. This is a lesson that the young schoolmaster in Goethe's *Elective Affinities* attempted to teach Charlotte, one of the married pair planning the garden, but to little effect.

39. Alexander Nehamas, *Nietzsche: Life as Literature*, 191–99; for a critique of Nehamas's further assumption that the appropriate literary model is a totally integrated organic unity—with implications for what it is to give style to one's life—see Shapiro, *Nietzschean Narratives*, 24, 86–89.

40. Kant, *Critique of Judgment*, section 22. Kant had reservations about the tendency of the picturesque English garden toward excessive complexity; like the baroque taste in furniture, he says "it carries the imagination's freedom very far, even to the verge of the grotesque."

41. In notes from the time of *HAH*, Nietzsche exclaims, "Waves—lapping at the shore on a calm summer day—Epicurus's garden-happiness" (*KSA* 8.527).

42. The famous tombstone inscription "*et in Arcadia ego*" that appears in two of Poussin's paintings known by this name has been subject to a variety of readings; in Nietzsche's aphorism, as in the epigraph to Goethe's *Italian Journey*, it apparently indicates nostalgia for a fragile beauty. The classic study of the iconology of the two Poussin paintings is Erwin Panofsky, "Et in Arcadia Ego: Poussin and the Elegiac Tradition," in *Meaning and the Visual Arts*, 295–320. For a useful account of the importance of landscape painting, especially that of Poussin and Claude Lorrain for landscape gardening, see Crandell, *Nature Pictorialized.*

43. For an exception, see Max Weber, who in 1904 wrote of the industrial capitalist compulsion to expand the exploitation of the earth "until the last ton of fossil fuel has burnt to ashes," *The Protestant Ethic and the Spirit of Capitalism*, 181.

44. Cf. Krell, *Postponements*; Babich, "Becoming and Purification: Empedocles, Zarathustra's *Übermensch*, and Lucian's Tyrant," in Lemm, ed., *Nietzsche and the Becoming of Life*, 245–61.

45. Jean-Luc Nancy, *After Fukushima: The Equivalence of Catastrophes.*

CHAPTER SIX

1. Giorgio Agamben, *Leviathan's Riddle*, 60.

2. Derrida, *The Politics of Friendship*, 26–48.

3. There is just one more literary reference by my count in *The Day After Tomorrow*, and it is to Nietzsche. Deciding which books to burn for warmth, one of the male high school students says his book must be saved because he is the greatest philosopher of the nineteenth century. The female student replies that he's a chauvinist pig who was in love with his sister.

4. When Nietzsche was eighteen he wrote in an essay, "That God became human refers only to this: that humans should seek their salvation not in eternity, but establish their heaven on earth instead"; "On Christianity," in *Young Nietzsche and Philosophy.*

5. Thomas Hobbes, *Leviathan*, chap. XXXV.

6. Schmitt, *Political Theology*, 36.

7. Giorgio Agamben, *The Kingdom and the Glory.*

8. Hegel, *Philosophy of World History*, 342.

9. Ibid., 442.

10. In *Anti-Nietzsche*, Malcolm Bull proposes the possibility of an indefinitely extended and continuing leveling-down process, such that nihilism would never reach the point of self-destruction or self-overcoming that Nietzsche assumes it will. Although this may be possible as a thought-experiment, what could make it desirable?

11. The Christian New Testament speaks at various times of the *kosmos*, sometimes as the whole of what is, sometimes as the "worldly" opposed to faith (as frequently in Matthew); *oikoumene* appears as simply designating the (known) inhabited world (e.g., Romans 10:18).

12. Yet an aggressively materialist philosopher like Hobbes could argue that even the final divine kingdom will be an earthly one.

13. Loeb, *The Death of Nietzsche's Zarathustra*, 45–84.

14. In addition to his documented knowledge of Zoroastrianism (from his readings in Hellwald and other scholars), Nietzsche was almost certainly familiar with von Hammer's book on *The History of the Assassins*, which argues (if rather tendentiously) that there is a lineage of thought and power from Zoroaster through the medieval Assassins, who ruled a significant swath of territory from their mountain stronghold in Alamut (in Persia, now northwestern Iran). In von Hammer's account, the Assassins transmitted their atheism, materialism, and communism to secret societies like the Freemasons. This widely read nineteenth-century book was very probably the source of the motto Nietzsche attributes to the Assassins, "Nothing is true, everything is permitted," and of the similar saying in Dostoyevsky's *The Brothers Karamazov* (*GM* III.24, *Z* IV.9). Cf. Shapiro, "Assassins and Crusaders: Nietzsche After 9/11."

The geography of the Assassins may have contributed to Nietzsche's imaginative belief in his descent from Polish nobility. See 202 n14.

15. He later names this virtue, identifying it as the gift-giving virtue (*die schenkende Tugend*) (*Z* III.10.2).

16. See, e.g., Daniel Conway, *Nietzsche's Dangerous Game,* 178–86.

17. Walter Kaufmann, trans. and ed., *The Portable Nietzsche*, 565.

18. See Jörg Salaquarda, "*Der Antichrist*" for a full discussion of the linguistic ambiguity, and Nietzsche's many uses of the term and its variations.

19. Walter Kaufmann, *Nietzsche: Philosopher, Psychologist, Antichrist.*

20. Kaufmann, *Nietzsche: Philosopher, Psychologist, Antichrist*, 29–30.

21. Sommer, *Friedrich Nietzsche's "Der Antichrist,"* 51–54. Salaquarda supports his understanding of Nietzsche's use of the Antichrist name or title as identical with Schopenhauer's concept of the immoralist by reference especially to his 1886 preface to the *Birth of Tragedy* (originally 1872) and his preface to the *Genealogy*. This is plausible as far as it goes, that is, with reference to those specific uses. Salaquarda wants to show that Nietzsche is not simply a resentful, reactive critic of Christianity, but that by adopting the title Antichrist he is taking a positive position that affirms this life and (only consequently) disdains Christianity. However, Sommer's commentary is strategically limited to the text of Nietzsche's *The Antichrist*. Here, as he maintains, the main burden of the text is indeed its opposition to and denunciation of Christianity, that is, the historical religion whose initiation Nietzsche credits to Paul.

22. An important book on Nietzsche's hero Frederick II is Ernst Kantorowicz's markedly Nietzschean study, written in the atmosphere of the Stefan Georg circle: *Frederick the Second, 1194–1250*.

23. See the thorough survey in Bernard McGinn, *Antichrist*.

24. McGinn speaks of the period 1660–1900 as "Antichrist in decline," *Antichrist*, 231–49.

25. Cf. Shapiro, "Nietzsche Contra Renan."

26. See Thomas Brobjer, "The Place and Role of *Der Antichrist* in Nietzsche's Four Volume Project *Umwertung aller Werte*."

27. See Christiane Koszka, "MELAS (Mitochondriale Enzephalomyopathie, Laktazidose und Schlaganfall-ähnliche Episoden)—eine neue Diagnose von Nietzsches Krankheit"; and critical responses by Roland Schiffter and Thomas Klopstock, in *Nietzsche-Studien* 42.

28. Andreas Sommer hypothesizes that almost the entire argument of *AC* can be read as a parody of Christian thinking (Sommer 2000, 688). Of course in Christian theology the Antichrist will indeed be a parody of Christ, as is clear as early as Hippolytus's treatise *On Christ and Antichrist*.

29. According to Gilles Quispel, "the Church Fathers, in their polemics against heresy, expressed for the first time the idea that there exists a development in history, the idea that in the education of the human race certain forms were justified in their time only to be rejected by a later epoch"; "Time and History in Patristic Christianity," in *Man and Time*, Joseph Campbell, ed., 88.

30. The *Refutation* is a crucial source of what we think of as the fragments of pre-Socratic philosophy and contains considerable testimony about them. It seems worth asking whether and to what extent some typical post-1850s pictures of the entire "development" from Thales to Socrates are structured in some ways by Hippolytus's program, but that is a question for another occasion.

31. Agamben, *The Time That Remains*, 15.

32. Carl Schmitt, *The Nomos of the Earth*, 59–60; Ernst Kantorowicz, *The King's Two Bodies*, 292; Giorgio Agamben, *The Kingdom and the Glory* 7, 15–16.

33. The early Cold War doctrine of "containment" of the Soviet Union was formulated by George F. Kennan, who wrote in an influential (but at first anonymous) 1947 article in *Foreign Affairs* that "the main element of any United States policy toward the Soviet Union must be a long-term, patient but firm and vigilant containment of Russian expansive tendencies. . . . Soviet pressure against the free institutions of the Western world is something that can be contained by the adroit and vigilant application of counterforce at a series of constantly shifting geographical and political points, corresponding to the shifts and manoeuvers of Soviet policy, but which cannot be charmed or talked out of existence." But what happens when "liberation," the policy of Eisenhower's Secretary of State John Foster Dulles and of Ronald Reagan replaces "containment," finally resulting in the symbolic fall of the Berlin Wall and the collapse of the Soviet Union? As Leonard Cohen expresses it in his song "The Future," it may mean the end of all divisions and principles of measure, something that leads him to beg, "Give me back the Berlin Wall, give me Stalin and St. Paul. I've seen the future, brother, and it's murder." Cf. *KSA* 13.368, where Nietzsche speaks of the mediocre as "the *Verzögerer* (delayers) par excellence."

34. Tertullian, *The Apology of Tertullian*, section XXXII. Similar observations are made by Jerome in his commentary on Daniel 7:8, and more of the same are attributed to Tyconius (d. ca. 390) in his *Commentary on Apocalypse*, a lost text, but one much cited, e.g., by Augustine.

35. *Nomos*, 59–60. Schmitt's post-World War II *The Nomos of the Earth*, with its articulation of the possibility of a global world order in the age of air power and its making explicit the katechontic dimension of political theology can be understood as sharing a problematic with what Nietzsche calls "a philosophy of the Antichrist."

36. English translation based on Giorgio Agamben's Italian translation. Agamben, *The Time That Remains*, 23.

37. Agamben, *The Time That Remains*, 1.

38. Ibid., 2.

39. Löwith, *From Hegel to Nietzsche*, 385.

40. It is tempting to think of Nietzsche and Overbeck as a kind of complementary yin and yang pair. Nietzsche was a flamboyant thinker who courted public discussion and opposition. He was proud of "having entered society through a duel," with his polemic against D. F. Strauss. Overbeck was modest and reserved; while he undermined the canon of Christian scripture by revealing falsifications and distortions, revealing the political motivations of canon formation, he did so quietly and without fanfare. While he published relatively little in his lifetime as explicit as his 1873 polemic, he left an enormous *Nachlass* of notes, memoirs, and essays, very little of which was available until recently. In the 1990s some of this work was published, including a fair sample of his *Kirchenlexicon*, a massive cross-referenced personal encyclopedia consisting of 20,000 octavo pages. Previously, an edition of some of these writings appeared as *Christentum und Kultur. Gedanken und Aufzeichnungen zur modernen Theologie*, edited by Overbeck's student Carl Bernoulli. Bernoulli also produced the two-volume, heavily documented *Franz Overbeck und Friedrich Nietzsche, eine Freundschaft* (1908).

Since the 1990s some German scholars, especially Andreas Sommer, have begun to consider the Nietzsche-Overbeck relation in a philosophical rather than biographical way. In the conflict between Nietzsche's "Basel" and "Weimar" friends, relatives, and associates, Overbeck was overshadowed and considerably outlived by Nietzsche's sister Elisabeth, another factor that contributed to downplaying the relationship and producing the conventional picture of Overbeck as simply a "friend" who helped Nietzsche with financial and other practical matters, finally fetching him from Turin after his collapse. See Sommer, *Der Geist der Historie und der Ende des Christentums,* and *Friedrich Nietzsche's "Der Antichrist."* For important evaluations, see Karl Löwith's conclusion to *From Hegel to Nietzsche*, 374–85, and Jacob Taubes, "The Demystification of Theology: Toward a Portrait of Overbeck," in *From Cult to Culture*, 147–61. For a good introduction to Overbeck in his relation to Nietzsche, see Lionel Gossman, *Basel in the Age of Burckhardt: A Study in Unseasonable Ideas*, 413–38.

41. See Taubes, *From Cult to Culture*, 157. As Taubes explains Overbeck's position: "History is not to be restrained by means of Christianity. 'Every attempt made to take the Christian periodization of history seriously' must shatter on this fact. The Christian reckoning of time would only be substantiated if Christianity had brought about 'a new era.' But Overbeck denies precisely this, for 'Christianity itself originally spoke of a new time under the prerequisite—one that was not met—that the existing world should perish and make room for a new one. This was,

for a brief moment a serious expectation and, as such an expectation, surfaced continually but fleetingly, never becoming a fact of historical permanence—which alone could have provided the real basis for an incontrovertible account of time, one corresponding to the facts of reality. The world, and not the Christian expectation of it, is what held its own.' "

42. Oehler and Bernoulli, eds., *Nietzsches Briefwechsel mit Overbeck*, 352. Overbeck wrote Nietzsche, reporting that Eberhard Vischer "has made and published a very beautiful scientific discovery: the Johannine Apocalypse is a Jewish text that has been given a Christian rewriting" (Dec 12, 1886, 86).

43. Overbeck, *Werke und Nachlass* 9:319–49.

44. On the Gnostic understanding of time, see Henri-Charles Puech, "Gnosis and Time," in Joseph Campbell, ed., *Man and Time*, 38–84, esp. 80–83.

45. Taubes, "The Demystification of Theology," in *From Cult to Culture,* 147.

46. Franz Overbeck, *How Christian Is Our Present-Day Theology?*

47. Quoted by Taubes, "The Demystification of Theology," in *From Cult to Culture,* 154.

48. Overbeck, *Christentum und Kultur, Werke* 6/1, 246. For Overbeck's 1873 critique of German theologians supporting Bismarck by arguments drawn from political theology, see *How Christian Is Our Present-Day Theology,* 63–67.

49. Overbeck, *Werke* 4, 157.

50. Franz Overbeck, *Quaestionum Hippolytearum* specimen.

51. A very strong version of the "Platonism for the people"—or more precisely Stoic universalism—thesis is found in Bruno Bauer's eccentric 1877 book *Christ and the Caesars*, reviewed by Overbeck in 1878. Nietzsche and Overbeck were corresponding frequently about theological matters around this time, and Nietzsche mentions Bauer in several contexts: *KSA* 6.317; *KSB* 6.242, 7.270, 275; 8.106, 205, 247, 370.

52. The multiplicity of regimes yet to come is compatible with Hippolytus's relatively anti-Roman stance, which can be contrasted with Tertullian's contemporary idea of Rome as *katechon*.

53. As Graham Parkes points out in his note on this passage, *duldet* is Luther's translation of the passage in I Cor. 13:7, where Paul writes of love or charity that it "endureth all things"; *Thus Spoke Zarathustra*, trans. Graham Parkes, 315.

54. Nietzsche to Overbeck, June 22, 1880; Overbeck to Nietzsche, July10, 1880, in Oehler and Bernoulli, eds., *Friedrich Nietzsches Briefwechsel mit Franz Overbeck*, 133–36.

55. Franz Overbeck, "Introduction to *The Acts of the Apostles*," 23–24 (accessible at http://archive.org/stream/contentsoriginof01zell#page/n17/mode/2up)

56. Overbeck, *Werke,* 392.

57. Hebrews is framed differently than those letters; it is not addressed to a specific group in a geographical location, and there is something odd about the supposed author, Paul, as apostle to the gentiles, writing to the Jews. As its title suggests, the text firmly links messianic faith to the Jewish tradition, citing sacred history and prophecy. The letter was accepted quickly in the Eastern Church, but was suspect for some time in the West; under dispute were such theological issues as the rejection of rebaptism in Hebrews 6:4–6 as well as philological questions.

58. See Taubes, *The Political Theology of Paul*, 4, for Taubes's anecdote concerning his conversation with the literary scholar Emil Staiger in Zürich during World War II. Staiger had been

reading Paul's epistles, but exclaimed bitterly: "But that isn't Greek, it's Yiddish! Upon which [Taubes] said: Yes, Professor, and that's why I understand it!"

59. Here Nietzsche's usage does seem to permit reading *Antichrist* as simply anti-Christian.

60. Karl Löwith, *Meaning in History*; Jacob Taubes, *Occidental Eschatology*; Agamben, *The Time That Remains*, 62–63.

61. Agamben, "Time and History: Critique of the Instant and the Continuum" in *Infancy and History: On the Destruction of Experience*, 97–116.

62. Seneca, *Letters to Lucilius*, 78, 14. Quoted by Pierre Hadot in *Philosophy as a Way of Life*, 228.

63. Shapiro, "The Text as Graffito: Historical Semiotics," in *Nietzschean Narratives*, 124–41.

64. Deleuze, *Difference and Repetition*, 86.

65. Ibid., 89.

66. It was found in Nietzsche's papers, glued to the last pages of *The Antichrist*; Colli and Montinari argue convincingly that it was intended to be the last page of that work (*KSA* 14.448–54).

67. In the Jewish and Christian imaginary, which Nietzsche probably shared on this point, there is a tendency to think of early Islam as defined by its wars. But the Islamic calendar begins with the Hijra, understood as an act of complete submission: Muhammad saves the faith and its revelation by leaving Mecca for Medina. Since Nietzsche praises the Islamic war on Christianity, he may have felt some kinship with what he took to be its associated way of dividing time into two parts. And like Jewish and Islamic dates, when presented in Christian or secular contexts, Nietzsche gives the other time scheme as a point of reference.

BIBLIOGRAPHY

Acampora, Christa. "Demos Agonistses Redux: Reflections on the Streit of Political Agonism." *Nietzsche-Studien* 32 (2003): 373–89.

———. *Contesting Nietzsche*. Chicago: University of Chicago Press, 2013.

Acampora, Christa, and Keith Ansell Pearson. *Nietzsche's Beyond Good and Evil*. London: Continuum, 2011.

Agamben, Giorgio. *Infancy and History: On the Destruction of Experience*. Translated by Liz Heron. New York: Verso, 2007.

———. *The Kingdom and the Glory*. Translated by Lorenzo Chiesa. Stanford: Stanford University Press, 2011.

———. *Language and Death: The Place of Negativity*. Translated by K. Pinkus and M. Hardt. Minneapolis: University of Minnesota Press, 1991.

———. *Leviathans Rätsel/Leviathan's Riddle*. Translated by Paul Silas Peterson. Tübingen: Mohr Siebeck, 2014.

———. *State of Exception*. Translated by Kevin Attell. Chicago: University of Chicago Press, 2005.

———. *The Time That Remains*. Translated by Patricia Dailey. Stanford: Stanford University Press, 2005.

Allison, David B., ed. *The New Nietzsche*. Cambridge: MIT Press, 1988.

Ansell-Pearson, Keith, ed. *A Companion to Nietzsche*. New York: Blackwell, 2006.

———. *An Introduction to Nietzsche as Political Thinker*. Cambridge: Cambridge University Press, 1994.

———, ed. *Nietzsche and Political Thought*. New York: Bloomsbury, 2014.

———. *Viroid Life*. New York: Routledge, 1997.

Arendt, Hannah. *The Origins of Totalitarianism*. New York: Harcourt Inc., 1979.

Bachelard, Gaston. *Air and Dreams*. Translated by E. and C. Farrell. Dallas: Dallas Institute Publications, Dallas Institute of Humanities and Culture, c1988.

———. *The Poetics of Space*. Translated by Maria Jolas. New York: Orion, 1964.

Badiou, Alain. *Being and Event*. Translated by Oliver Feltham. New York: Continuum, 2005.

———. *Casser en deux l'histoire du monde?* Paris, 1992.

———. *Infinite Thought*. Translated by Oliver Feltham and James Clemens. New York: Continuum, c. 2008.

———. *Manifesto for Philosophy*. Translated by Norman Madarasz. Albany: State University of New York Press, 1999.

———. "Who Is Nietzsche?" *Pli* 11 (2001): 1–11.

Balibar, Étienne. "Spinoza, the Anti-Orwell: Fear of the Masses." *Rethinking Marxism*, 2 (3) (1989): 104–39.

Bauer, Bruno. *Christus und die Caesaren: der Ursprung der Christentum aus dem römischen Griechenthum*. Hildesheim: G. Olms, 1968.

——— (anonymous publication). *The Trumpet of the Last Judgment Against Hegel, the Atheist and Antichrist*. Translated by Lawrence Stepelevich. Lewiston, NY: Edwin Mellen, 1989.

Beehler, Sharon. "'Confederate Season': Shakespeare and the Elizabethan Understanding of *Kairos*." In *Shakespeare Matters*, edited by Lloyd Davis, 74–88. Newark: University of Delaware Press, 2003.

Bergmann, Peter. *Nietzsche: The Last Antipolitical German*. Bloomington: Indiana University Press, 1987.

Bernoulli, Carl. *Franz Overbeck und Friedrich Nietzsche, eine Freundschaft*. 2 vols. Jena: E. Diederichs, 1908.

Bertram, Ernst. *Nietzsche: Attempt at a Mythology*. Robert E. Norton. Urbana, University of Illinois Press, 2009.

Blaise, Clark. *Time Lord: Sir Sandford Fleming and the Creation of Standard Time*. New York: Random House, 2002.

Böhme, Gernot. *Atmosphäre: Essays zur neuen Ästhetik*. Berlin: Suhrkamp, 2013.

Bosteels, Bruno. "Nietzsche, Badiou, and Grand Politics: An Antiphilosophical Reading." In *Nietzsche and Political Thought*, edited by Keith Ansell-Pearson, 213–39. London: Bloomsbury, 2013.

Brobjer, Thomas H. "Notes and Discussions: Nietzsche's Knowledge of Kierkegaard." *Journal of the History of Philosophy* 41, no. 2 (2003): 251–63.

———. "The Place and Role of *Der Antichrist* in Nietzsche's Four Volume Project *Umwertung aller Werte*." *Nietzsche-Studien* 40 (2011): 244–55.

Buddensieg, Tilman. *Nietzsches Italien: Städte, Gärten, Paläste*. Berlin: Verlag Klaus Wagenbach, 2002.

Bühler, Pierre. "Tertullian: The Teacher of the Credo Quia Absurdum." In *Kierkegaard and the Patristic and Medieval Traditions*, edited by Jon Stewart, 131–42. Burlington: Ashgate, 2008.

Bull, Malcolm. *Anti-Nietzsche*. Brooklyn: Verso, 2011.

Burckhardt, Jacob. *Der Cicerone: Eine Anleitung zum Genuss der Kunstwerke Italiens*. Stuttgart: Alfred Kröner, 1964.

———. *Reflections on History*. Translated by M.D. Hottinger. Indianapolis: Liberty Classics, 1979.

Burnham, Douglas. *Reading Nietzsche: An Analysis of Beyond Good and Evil.* Montreal: McGill-Queen's University Press, 2007.

Cameron, Frank, and Don Dombowsky, eds. *Political Writings of Friedrich Nietzsche.* New York: Palgrave Macmillan, 2008.

Campbell, Joseph, ed. *Man and Time: Papers from the Eranos Yearbook.* Princeton: Princeton University Press, 1973.

Cancik, Hubert. "'Mongols, Semites, and the Pure-bred Greeks': Nietzsche's Handling of the Racial Doctrines of His Time." In *Nietzsche and Jewish Culture*, edited by J. Golomb, 55–75. New York: Routledge, 1997.

Casey, Edward S. *Getting Back into Place: Toward a Renewed Understanding of the Place-World.* Bloomington: Indiana University Press, 1993.

Cavell, Stanley. *Conditions Handsome and Unhandsome.* Chicago: University of Chicago Press, 1990.

———. *This New Yet Unapproachable America: Lectures after Emerson after Wittgenstein.* Albuquerque: Living Batch Press, 1989.

Citron, Bernard. "The Multitude in the Synoptic Gospels." *Scottish Journal of Theology* 7:4 (December 1954), 408–18.

Conant, James. "Nietzsche's Perfectionism: A Reading of Schopenhauer as Educator." In *Nietzsche's Postmoralism*, edited by Richard Schacht, 181–257. Berkeley: University of California Press, 2001.

Conjonctures, 45–46 (2008). "Nietzsche et la multitude."

Conway, Daniel. *Nietzsche and the Political.* New York: Routledge, 1997.

———. *Nietzsche's Dangerous Game.* New York: Cambridge University Press, 1997.

———, ed. With Peter S. Groff. *Nietzsche: Critical Assessments*, 4 vols. New York: Routledge, 1998.

Crandell, Gina. *Nature Pictorialized: "The View" In Landscape History.* Baltimore: Johns Hopkins University Press, 1993.

Critchley, Simon. *Infinitely Demanding: Ethics of Commitment, Politics of Resistance.* New York: Verso, 2007.

Del Caro, Adrian. *Grounding the Nietzsche Rhetoric of Earth.* New York: de Gruyter, 2004.

Deleuze, Gilles. *Difference and Repetition.* Translated by P. Patton. New York: Columbia University Press, 1994.

———. *Essays Critical and Clinical.* Translated by Daniel W. Smith and Michael A. Greco. Minneapolis: University of Minnesota Press, 1997.

———. *Foucault.* Translated by Seán Hand. Minneapolis: University of Minnesota Press, 1988.

———. "Nomad Thought." Translated by David B. Allison. In *The New Nietzsche,* edited by David B. Allison, 142–49. Cambridge: MIT Press, 1988.

Deleuze, Gilles, and Felix Guattari. *Anti-Oedipus: Capitalism and Schizophrenia.* Translated by Robert Hurley et al. New York: Penguin, 2009.

———. *A Thousand Plateaus.* Translated by Brian Massami. Minneapolis: University of Minnesota Press, 1988.

———. *What Is Philosophy?* Translated by Hugh Tomlinson and Graham Burchell. New York: Columbia University Press, 1994.

Deleuze, Gilles, and Claire Parnet. *Dialogues*. Translated by Hugh Tomlinson and Barbara Habberjam. New York: Columbia University Press, 1987.

Derrida, Jacques. *The Death Penalty: Vol. 1*. Translated by Peggy Kamuf. Chicago: University of Chicago Press, 2014.

———. *The Politics of Friendship*. Translated by George Collins. New York: Verso, 1997.

———. *The Truth in Painting*. Translated by Geoff Bennington. Chicago: University of Chicago Press, 1987.

Deussen, Paul. *The System of the Vedanta*. Translated by Charles Johnson. New York: Dover, 1973.

Emerson, Ralph W. "History." In *Essays First and Second Series*, 5–26. New York: Vintage Books/The Library of America, 1990.

Figal, Günter. Lecture at the Collegium Philosophicum, July 2001.

Figl, Johann. *Nietzsche und die Religionen: Transkulturelle Perspektiven seines Bildungs- und Denkweges*. Berlin: de Gruyter, 2007.

Foucault, Michel. *Discipline and Punish: The Birth of the Prison*. Translated by Alan Sheridan. New York: Vintage, 1979.

———. *The History of Sexuality: Vol. 1*. Translated by Robert Hurley. New York: Vintage, 1990.

———. "La force de fuir." In *Dits et écrits*, edited by Daniel Defert and François Ewald, vol. 2, 401–05. Paris: Editions Gallimard, 1994.

———. "Nietzsche, Genealogy, History." In *Language, Counter-Memory, Practice*, edited by Donald F. Bouchard, 139–64. Ithaca: Cornell University Press, 1977.

Friedrich, Carl. *Constitutional Reason of State: The Survival of the Constitutional Order*. Providence: Brown University Press, 1957.

Fukuyama, Francis. *The End of History and the Last Man*. New York: Free Press, 1992.

Gilman, Sander, ed. *Conversations with Nietzsche*. New York: Oxford University Press, 1987.

Goethe, Johann Wolfgang. *Elective Affinities*. Translated by Victor Lange and Judith Ryan. New York: Suhrkamp, 1988.

———. *Faust*. Translated by Walter Kaufmann. New York: Doubleday, 1961.

Golomb, Jacob, ed. *Nietzsche and Jewish Culture*. New York: Routledge, 1997.

Golomb, Jacob, and Robert Wistrich, eds., *Nietzsche, Godfather of Fascism?* Princeton: Princeton University Press, 2002.

Gooding-Williams, Robert. *Zarathustra's Dionysian Modernism*. Stanford: Stanford University Press, 2001.

Gossman, Lionel. *Basel in the Age of Burckhardt: A Study in Unseasonable Ideas*. Chicago: University of Chicago Press, 2000.

Graeber, David. *Debt: The First Five Thousand Years*. Brooklyn: Melville House, 2011.

Grimm, Jacob and Wilhelm. *Deutsches Wörterbuch*: Leipzig: S. Hirzel, 1854.

Günzel, Stephan. *Geophilosophie: Nietzsches philosophische Geographie*. Berlin: Akademie Verlag, 2001.

Hadot, Pierre. *Philosophy as a Way of Life*. Edited by Arnold Davidson. New York: Blackwell, 1995.

Hardt, Michael, and Antonio Negri. *Empire*. Cambridge: Harvard University Press, 2000.

———. *Multitude: War and Democracy in the Age of Empire*. New York: Penguin, 2005.

Harris, Horton. *David Friedrich Strauss and His Theology*. Cambridge: Cambridge University Press, 1973.

Hartmann, Eduard von. *Philosophy of the Unconscious*. Translated by W. C. Coupland. London: Kegan Paul, 1931.

Hatab, Lawrence. *A Nietzschean Defense of Democracy*. Chicago: Open Court, 1995.

Hegel, Georg W. F. *Hegel's Aesthetics*. Translated by T. M. Knox. Oxford: Oxford University Press, 1975.

———. *Introduction to the Philosophy of History*. Translated by Leo Rauch. Indianapolis: Hackett, 1988.

———. *The Logic of Hegel*. (*Encyclopedia I*) Translated by W. Wallace. New York: Oxford University Press, 1959.

———. *The Phenomenology of Spirit*. Translated by A.V. Miller. New York: Oxford University Press, 1977.

———. *Philosophy of History*. Translated by J. Sibree. New York: Dover, 1955.

———. *Philosophy of Mind*. (*Encyclopedia III*) Translated by William Wallace. New York: Oxford University Press, 1971.

———. *Philosophy of Right*. Translated by T. M. Knox. Oxford: Oxford University Press, 1958.

Heidegger, Martin. "Anaximander's Saying." In *Off the Beaten Track*, edited and translated by Julian Young and Kenneth Haynes, 242–81. New York: Cambridge University Press, 2002.

———. *Being and Time*. Translated by Joan Stambaugh. Albany: State University of New York Press, 1996.

———. *Poetry, Language, Thought*. Translated by Albert Hofstadter. New York: Harper and Row, 1975.

Hell, Julia "Katechon: Carl Schmitt's Imperial Theology and the Ruins of the Future." *Germanic Review* 2009: 283–326.

Henry, Martin. *Franz Overbeck: Theologian?* New York: Peter Lang, 1995.

Hippolytus of Rome. *Refutation of All Heresies* and *Treatise on Christ and Antichrist*. In *The Anti-Nicene Fathers*, 9–162 and 204–19, edited by Alexander Roberts and James Donaldson. New York: Charles Scribner's Sons, 1926.

Hobbes, Thomas. *Leviathan*. Edited by Edwin Curley. Indianapolis: Hackett, 1994.

Hofstadter, Albert. *Agony and Epitaph*. New York: George Braziller, 1970.

Holborn, Hajo. *A History of Modern Germany: 1840–1945*. New York: Knopf, 1969.

Holub, Robert. "Reading Nietzsche as Postmodernist." In *Why Literature Matters: Theories and Functions of Literature*, edited by Ahrens Rüdiger and Laurence Volkmann, 247–63. Heidelberg: Universitäts Verlag C. Winter, 1996.

Hunt, John Dixon. "'Ut Pictura Poesis': The Garden and the Picturesque in England (1710–1750)." In *The History of Garden Design: The Western Tradition from the Renaissance to the Present Day*, edited by Monique Mosser and George Teyssot. London: Thames & Hudson, 1991.

Hunt, John Dixon, and Peter Wills. *The Genius of the Place: The English Landscape Garden 1620–1820*. Cambridge: The MIT Press, 1988.

Husserl, Edmund. "Foundational Investigations of the Phenomenological Origin of the Spatiality of Nature: The Originary Ark, the Earth, Does Not Move." In Maurice Merleau-Ponty,

Husserl at the Limits of Phenomenology, edited by Leonard Lawlor with Bettina Bergo, 117–31. Evanston: Northwestern University Press, 2002.

Irigaray, Luce. *Marine Lover of Friedrich Nietzsche*. Translated by Gillian Gill. New York: Columbia University Press, 1991.

Jacolliot, Louis. *Les législateurs religieux. Manou-Moïse-Mahomet*. Paris: A. Lacroix, 1876.

Jensen, Anthony K. "The Rogue of All Rogues: Nietzsche's Presentation of Eduard von Hartmann's *Philosophie des Unbewussten* and Hartmann's Response to Nietzsche." *The Journal of Nietzsche Studies* 32 (Autumn 2006): 41–61.

Jung, Carl. *Nietzsche's Zarathustra*, 2 vols. Edited by James Jarrett. Princeton: Princeton University Press, 1988.

Kahn, Charles. *"Anaximander's Fragment: The Universe Governed by Law."* In *The Pre-Socratics: A Collection of Critical Essays,* edited by Alexander P. D. Mourelatos, 99–117. New York: Anchor, 1974.

Kant, Immanuel. *Critique of Judgment*. Translated by Werner Pluhar. Indianapolis: Hackett, 1986.

———. *Critique of Pure Reason*. Translated by Werner Pluhar. Indianapolis: Hackett, 1996.

Kantorowicz, Ernst. *Frederick II, 1194–1250*. Translated by E. O. Lorimer. New York: Ungar, 1967.

———. *The King's Two Bodies*. Princeton: Princeton University Press, 1985.

Kaufmann, Walter. *Nietzsche: Philosopher, Psychologist, Antichrist*, 4th ed. Princeton: Princeton University Press, 1974.

———, trans. and ed. *The Portable Nietzsche*. New York: Viking, 1954.

[Kennan, George F.]. Writing as Anonymous. "The Sources of Soviet Conduct." *Foreign Affairs* 25 (July 1947), 566–82.

Kierkegaard, Søren. *Concluding Unscientific Postscript to the Philosophical Crumbs*. Translated by Alastair Hannay. New York: Cambridge University Press, 2009.

Klopstock, Thomas. "Friedrich Nietzsche und seine Krankheiten: kein ausreichender Anhalt für MELAS." *Nietzsche-Studien* 42 (2013): 293–97.

Kojève, Alexandre. *Introduction to the Reading of Hegel*. Translated by James H. Nichols, Jr. New York: Basic Books, 1969.

Kostka, Alexandre, and Irving Wohlfarth, eds. *Nietzsche and "An Architecture of Our Minds."* Los Angeles: Getty Research Institute, 1999.

Koszka, Christiane. "MELAS (Mitochondriale Enzephalomyopathie, Laktazidose und Schlaganfall-ähnliche Episoden)—eine neue Diagnose von Nietzsches Krankheit." *Nietzsche-Studien* 39 (2010): 573–78.

Krell, David Farrell. *Postponements: Woman, Sensuality, and Death in Nietzsche*. Bloomington: Indiana University Press, 1986.

Kristeller, Paul O. "The Modern System of the Arts." In *Renaissance Thought II*, 163–227. New York: Harper & Row, 1965.

Lampert, Lawrence. *Leo Strauss and Nietzsche*. Chicago: University of Chicago Press, 1996.

———. "Nietzsche's Philosophy and True Religion." In *A Companion to Nietzsche*, edited by Keith Ansell-Pearson, 133–47. New York: Blackwell, 2006.

———. *Nietzsche's Task: An Interpretation of Beyond Good and Evil*. New Haven: Yale University Press, 2001.

———. *Nietzsche's Teaching: An Interpretation of Thus Spoke Zarathustra*. New Haven: Yale University Press, 1986.

Lange, Frederick Albert. *The History of Materialism*. Translated by Ernest Chester Thomas. New York: Harcourt, Brace and Company, 1925.

Lawrence, David Herbert. *Apocalypse and the Writings on Revelation*. Edited by Mara Kalnins. New York: Cambridge University Press, 2002.

Lefebvre, Henri. *Production of Space*. Translated by Donald Nicholson-Smith. Cambridge: Blackwell, 1991.

Lemm, Vanessa, ed. *Nietzsche and the Becoming of Life*. New York: Fordham, 2015.

———. "Nietzsche's Great Politics of the Event." in *Nietzsche and Political Thought*, edited by Keith Ansell-Pearson, 179–96. London: Bloomsbury, 2013.

Lessing, Gotthold Ephraim. *Emilia Galotti*. Translated by E. Dvoretzky. New York: Ungar, 1962.

Lippit, John, ed. *Nietzsche's Futures*. New York: St. Martin's Press, 1999.

Loeb, Paul. *The Death of Nietzsche's Zarathustra*. New York: Cambridge University Press, 2010.

Löwith, Karl. *From Hegel to Nietzsche: The Revolution in Nineteenth-Century Thought*. Translated by David E. Green. New York: Anchor, 1967.

———. *Meaning in History*. Chicago: University of Chicago Press, 1960.

Machiavelli, Niccolo. *The Prince*. Translated by Robert Adams. New York: W. W. Norton & Company, 1977.

Magnus, Bernd, and Kathleen Higgins, eds. *The Cambridge Companion to Nietzsche*. New York: Cambridge University Press, 1996.

Malbon, Elizabeth. *In the Company of Jesus: Characters in Mark's Gospel*. Louisville: Westminster John Knox Press, 2000.

Marx, Karl. *Selected Writings*. Edited by Lawrence Simon. Indianapolis: Hackett, 1994.

McCumber, John. *Time in the Ditch*. Evanston: Northwestern University Press, 2001.

McGinn, Bernard. *Antichrist: Two Thousand Years of the Human Fascination with Evil*. New York: Harper, 2000.

———, ed. *Apocalyptic Spirituality*. New York: Paulist Press, 1979.

Merleau-Ponty, Maurice. *The Phenomenology of Perception*. Translated by Colin Smith. New York: Humanities Press, 1962.

Montinari, Mazzino. *Reading Nietzsche*. Urbana: University of Illinois, 2003.

Nancy, Jean-Luc. *After Fukushima: The Equivalence of Catastrophes*. Translated by Charlotte Mandell. New York, Fordham University Press, 2015.

———. *The Creation of the World, or, Globalization*. Translated by François Raffoul and David Pettigrew. Albany: State University of New York Press, 2007.

Negri, Antonio. *Time for Revolution*. Translated by Matteo Mandarini. London: Continuum, 2005.

Nehamas, Alexander. *Nietzsche: Life as Literature*. Cambridge: Harvard University Press, 1985.

Niemeyer, Christian, ed. *Nietzsche-Lexicon*. 2nd ed. Darmstadt: Wissenschaftliche Buchgesellschaft, 2011.

O'Brien, Dan, ed. *Gardening: Cultivating Wisdom*. Malden, MA: Blackwell, 2010.

Oehler, R., and Carl Bernoulli, eds. A. *Nietzsches Briefwechsel mit Overbeck*. Leipzig: Insel-Verlag, 1916.

Overbeck, Franz. *Christentum und Kultur. Gedanken und Aufzeichnungen zur modernen Theologie*. Edited by Carl Bernoulli. Basel: Benno Schwabe & Co., 1919. (Reprinted in Overbeck, *Werke und Nachlass*).

———. *How Christian Is Our Present-Day Theology?* Translated by Martin Henry. London: Continuum, 2005.

———. "Introduction to the Acts of the Apostles." In Eduard Zeller, *The Contents and Origins of the Acts of the Apostles*. Translated by Joseph Dare. London: Williams & Norgate, 1875, 23–24.

———. *Quaestionum Hippolytearum specimen*. [*De Hippolyti libello de Antichristo*.] Dissertation. Jena, 1864.

———. Werke und Nachlass. 9 vols. Edited by Marianne Stauffacher Schaub, Ekkehard W. Stegemann, Niklaus Peter et al. Stuttgart: J. B. Metzler, 1994–2000.

Panofsky, Erwin. "Et in Arcadia Ego: Poussin and the Elegiac Tradition." In *Meaning and the Visual Arts*. New York: Doubleday, 1955, 295–320.

———. *Studies in Iconology: Humanistic Themes in the Art of the Renaissance*. New York: Harper, 1962.

Parkes, Graham. "Staying Loyal to the Earth: Nietzsche as an Ecological Thinker." In *Nietzsche's Futures*, edited by John Lippit. New York: St. Martin's Press, 1999.

Passant, E. J. *A Short History of Germany: 1815–1945*. Cambridge: Cambridge University Press, 1969.

Porter, James. *Nietzsche and the Philology of the Future*. Stanford: Stanford University Press, 2000.

Price, Uvedale. *Essays on the Picturesque*. vol. I. London: J. Mawman, 1810.

Rancière Jacques. "Who Is the Subject of the Rights of Man?" *The South Atlantic Quarterly* 103: 2/3, Spring/Summer 2004, 297–310.

Ratzel, Friedrich. *Anthropo-Geographie*. Stuttgart: J. Engelhorm, 1882.

Richardson, John. *Nietzsche's New Darwinism*. New York: Oxford University Press, 2004.

Rogers, Elizabeth Barlow. *Landscape Design: A Cultural and Architectural History*. New York: Harry N. Abrams, 2001.

Rohl, J. C. G. *From Bismarck to Hitler*. London: Longman, 1970.

Rosenthal, Bernice Glatzer. *Nietzsche and Soviet Culture: Ally and Adversary*. New York: Cambridge University Press, 2010.

Rossiter, Clinton. *Constitutional Dictatorship: Crisis Government in the Modern Democracies*. New Brunswick: Transaction, 2002.

Salaquarda, Jörg. "*Der Antichrist*." *Nietzsche-Studien* 2: 1973, 95–136.

Schacht, Richard, ed. *Nietzsche's Postmoralism*. Berkeley: University of California Press, 2001.

Schiffter, Roland. "Friedrich Nietzsches Krankheiten—eine unendliche Geschichte." In *Nietzsche-Studien* 42 (2013): 283–92.

Schmitt, Carl. *The Nomos of the Earth in the International Law of the Jus Publicum Europaeum*. Translated by G. L. Ulmen. New York: Telos Press, 2003.

———. *Political Theology*. Translated by George Schwab. Chicago: University of Chicago Press, 2005.

———. *Staat, Grossraum, Nomos: Arbeiten aus den Jahren 1916–1969*. Edited by Günther Maschke. Berlin, Duncker & Humblot, 1995.

Schopenhauer, Arthur. *Parerga and Paralipomena*. 2 vols. Translated by E. F. J. Payne. Oxford: Clarendon Press, 1974.

———. *The World as Will and Representation*. Translated by E. F. J. Payne. New York: Dover, 1966.

Scott, Peter, and William Cavanaugh, eds. *The Blackwell Companion to Political Theology*. New York: Blackwell, 2004.

Scott, James. *Seeing Like a State: How Certain Schemes to Improve the Human Condition Have Failed*. New Haven: Yale University Press, 1998.

Shapiro, Gary. *Alcyone: Nietzsche on Gifts, Noise, and Women*. Albany: State University of New York Press, 1991.

———. *Archaeologies of Vision: Foucault and Nietzsche on Seeing and Saying*. Chicago: University of Chicago Press, 2003.

———. "Assassins and Crusaders: Nietzsche After 9/11." In *Nietzsche at the Margins*, edited by Alan Hicks and Alan Rosenberg, 186–204. Lafayette: Purdue University Press, 2007.

———. "Beyond Peoples and Fatherlands: Nietzsche's Geophilosophy and the Direction of the Earth." *Journal of Nietzsche Studies* 35 (2008): 9–27.

———. "Debts Due and Overdue: Beginnings of Philosophy in Nietzsche, Heidegger, and Anaximander." In *Nietzsche, Genealogy, Morality*, edited by Richard Schacht, 358–75. Berkeley: University of California Press, 1994.

———. "Dogs, Domestication, and the Ego." In *A Nietzschean Bestiary, edited by* Christa and Ralph Acampora, 53-60. Rowman and Littlefield: New York, 2003.

———. *Earthwards: Robert Smithson and Art after Babel*. Berkeley: University of California Press, 1995.

———. *Nietzschean Narratives*. Bloomington: Indiana University Press, 1989.

———. "Nietzsche Contra Renan." *History and Theory*, 1982: 2, 193–222.

———. "Nietzsche on Geophilosophy and Geoaesthetics." In *The Blackwell Companion to Nietzsche*, edited by Keith Ansell-Pearson 477–94. New York: Blackwell, 2006.

———. "Nietzsche's Modernity: On Robert Gooding-Williams's *Zarathustra's Dionysian Modernism*." *International Studies in Philosophy*, 2004: 99–115.

———. "Nietzsche's Unmodern Thinking: Globalization, the End of History, and 'Great Events.'" *American Catholic Philosophical Quarterly* 84:2 (Spring 2010): 205–30.

———. "Notes on the Animal Kingdom of the Spirit." *Clio* (Spring 1979), 323–38.

———. "States and Nomads: Hegel's World and Nietzsche's Earth." In *Nietzsche and the Becoming of Life*, edited by V. Lemm, 303–17. New York: Fordham University Press, 2015.

———. "Territories, Landscapes, Gardens: Toward Geoaesthetics." *Angelaki: Journal of the Theoretical Humanities* 9:2 (2004): 103–15.

———. "Translating, Repeating, Naming: Foucault Derrida, and *The Genealogy of Morals*." In *Nietzsche as Postmodernist: Essays Pro and Contra*, edited by Clayton Koelb, 39–56. Albany: State University of New York Press, 1990.

Shaw, Tamsin. *Nietzsche's Political Skepticism*. Princeton: Princeton University Press, 2007.

Shepard, Paul. *Man in the Landscape: A Historic View of the Esthetics of Nature*. Atlanta: University of Georgia Press, 2002.

Shiner, Larry. *The Invention of Art: A Cultural History*. Chicago: University of Chicago Press, 2001.

Siemens, Herman, and Vasti Roodt, eds. *Nietzsche, Power, and Politics: Rethinking Nietzsche's Legacy for Political Thought.* New York: de Gruyter, 2009.

Simon, W. M. *Germany in the Age of Bismarck.* London: George Allen and Unwin, 1968.

Sloterdijk, Peter. *Spheres I. Bubbles.* New York: Semiotext(e), 2011.

———. *Spheres II. Globes.* New York: Semiotext(e), 2014.

———. *Spheres III. Schäume.* Frankfurt: Suhrkamp, 2004.

———. *You Must Change Your Life.* Translated by Wieland Hoban. Malden, MA: Polity Press, 2013.

Smith, Morton. *Jesus the Magician.* San Francisco: Harper & Row, 1978.

Sommer, Andreas. *Der Geist der Historie und der Ende des Christentums. Zur "Waffengenossenheit" Friedrich Nietzsche und Franz Overbeck. Mit einen Anhang unpublitzierter Texte aus Overbecks "Kirchenlexicon."* Berlin: Akademie Verlag, 1997.

———. *Friedrich Nietzsche's "Der Antichrist": Ein philosophisch-historischer Kommentar.* Basel: Schwabe & Co., 2000.

Spengler, Oswald. *The Decline of the West.* 2vols. Translated by Charles Francis Atkinson. New York: Alfred A. Knopf, 1930.

Spivak, Gayatri Chakravorty. "Can the Subaltern Speak?" In *Marxism and the Interpretation of Culture*, edited by C. Nelson and L. Grossberg, 271–313. Urbana: University of Illinois Press, 1988.

Stack, George. *Nietzsche and Emerson: An Elective Affinity.* Athens: Ohio University Press, 1992.

Strauss, David Friedrich. *The Old Faith and the New.* Translated by Mathilde Blind. Edited by G. A. Wells. Amherst NY: Prometheus Press, 1997.

Strong, Tracy B. *Friedrich Nietzsche and the Politics of Transfiguration.* 3rd ed. Urbana: University of Illinois Press, 2000.

———, ed. *Friedrich Nietzsche*: International Library of Essays in the History of Social and Political Thought. Burlington, VT: Ashgate, 2009.

———. "Nietzsche's Political Misappropriation." In *The Cambridge Companion to Nietzsche*, edited by Bernd Magnus and Kathleen Higgins, 119–47. New York: Cambridge University Press, 1996.

Szendy, Peter. *Apocalypse-Cinema: 2012 and Other Ends of the World.* Translated by Will Bishop. New York: Fordham University Press, 2015.

Tang, Chenxi. *The Geographic Imagination of Modernity: Geography, Literature, and Philosophy in German Romanticism.* Stanford: Stanford University Press, 2008.

Taubes, Jacob. *From Cult to Culture: Fragments Toward a Critique of Historical Reason.* Edited by Charlotte Fonrobert and Amir Engel. Stanford: Stanford University Press, 2010.

———. *Occidental Eschatology.* Translated by D. Ratmoko. Stanford: Stanford University Press, 2009.

———. *The Political Theology of Paul.* Translated by Dana Hollander. Stanford: Stanford University Press, 2004.

Tertullian. *The Apology of Tertullian.* In Tertullian, *Apology, De Spectaculis.* vol. 250. Cambridge, Mass.: Harvard University Press; London: W. Heinemann, 1953.

Von Hammer, Joseph. *Die Geschichte der Assassinen aus Morgenländischen Quellen.* Stuttgart and Tübingen: Cotta'schen Buchhandlung, 1818.

Vermes, Geza. *Christian Beginnings*. New Haven: Yale University Press, 2012.

Warren, Mark. *Nietzsche and Political Thought*. Cambridge: MIT Press, 1988.

Weber, Max. *The Protestant Ethic and the Spirit of Capitalism*. Translated by Talcott Parsons. New York: Charles Scribner's Sons, 1958.

Wellhausen, Julius. *Prolegomena to the History of Israel*. Translated by J. Sutherland Black and Alan Menzies. Atlanta: Scholars Press, 1994.

Yovel, Yirmiyahu. "Nietzsche and the Jews: The Structure of an Ambivalence." In *Nietzsche and Jewish Culture*, edited by Jacob Golomb, 117–36. New York: Routledge, 1997.

Žižek, Slavoj. *Living in the End Times*. London: Verso, 2010.

———. *The Puppet and the Dwarf: The Perverse Core of Christianity*. Cambridge: MIT Press, 2003.

Žižek, Slavoj, and Boris Gunjević. *God in Pain: Inversions of Apocalypse*. New York: Seven Stories Press, 2012.

Zupančič, Alenka. *The Shortest Shadow: Nietzsche's Philosophy of the Two*. Cambridge: MIT Press, 2003.

INDEX